Brush & Weeds
of Texas Rangelands

AUTHORS

CHARLES R. HART — Professor, Associate Department Head and Extension Program Leader for Ecosystem Science and Management

BARRON RECTOR — Associate Professor and Extension Range Specialist

C. WAYNE HANSELKA — Professor and Extension Range Specialist

ROBERT K. LYONS — Professor and Extension Range Specialist

ALLAN McGINTY — Professor and Extension Range Specialist

EDITOR

DIANE BOWEN — Associate Editor and Extension Communications Specialist

DESIGNER

LORI COLVIN — Assistant Graphic Designer and Extension Communications Specialist

All of the Texas AgriLife Extension Service,
The Texas A&M University System

About the cover...

TEXAS THISTLE
*(C. texanum)
is one of the more common species of thistles
found in the Lone Star State.*

PHOTOGRAPH BY CHARLES R. HART

Plant photos throughout this book were contributed by:

C. Wayne Hanselka, Charles R. Hart, Robert K. Lyons,
Allan McGinty, John C. Reagor and Barron Rector

*Support for this publication was provided
in part by the USDA CSREES
Renewable Resources Extension Act.*

Improving Lives. Improving Texas.

Copyright © 2008 by the Texas AgriLife Extension Service
All rights reserved
ISBN 0-9721049-4-1

Contents

- i Preface
- iii Vegetational Areas of Texas
- 1 Plant Descriptions
- 2 African rue
- 4 Annual broomweed, common broomweed
- 6 Annual croton: One-seed croton
- 8 Annual croton: Texas croton
- 10 Annual croton: Woolly croton
- 12 Ashe juniper, blueberry juniper
- 14 Berlandier lobelia
- 16 Bigelow shinoak, scalybark oak, white shinoak
- 18 Bitter sneezeweed, basin sneezeweed
- 20 Blackberry, dewberry
- 22 Blackbrush
- 24 Blackjack oak
- 26 Bois d'arc, osage orange
- 28 Broom snakeweed, perennial broomweed
- 30 Buffalo-bur
- 32 Burrobrush
- 34 Camphorweed
- 36 Carolina horse nettle
- 38 Catclaw acacia
- 40 Catclaw mimosa, fragrant mimosa
- 42 Ceniza, purplesage, Texas silverleaf
- 44 Chinese tallow tree
- 46 Cholla
- 48 Christ thorn, Jerusalem thorn
- 50 Cocklebur
- 52 Common goldenweed, Drummonds goldenweed
- 54 Common persimmon, eastern persimmon
- 56 Common sunflower, annual sunflower
- 58 Creosotebush, greasewood
- 60 Curlycup gumweed
- 62 Dog cactus, dog cholla, clavellina
- 64 Eastern baccharis
- 66 Eastern red cedar
- 68 Elm
- 70 Flameleaf sumac
- 72 Garboncillo, rattleweed
- 74 Gray goldaster
- 76 Greenbriar, saw greenbriar
- 78 Guajillo
- 80 Hackberry
- 82 Hercules club, toothache tree
- 84 Honey locust
- 86 Honey mesquite
- 88 Horehound
- 90 Huisache
- 92 Lime pricklyash, colima
- 94 Lotebush, blue brush, gumdrop tree
- 96 Macartney rose
- 98 Marshelder, narrowleaf sumpweed
- 100 Mohrs shin oak
- 102 Plantain
- 104 Post oak
- 106 Prairie gerardia, prairie agalinis
- 108 Pricklyash, tickle-tongue
- 110 Pricklypear
- 112 Rayless goldenrod, jimmyweed

(continued on next page)

Contents (continued)

114	Redberry juniper, Pinchot juniper		154	Texas persimmon
116	Retama		156	Thistles
118	Roosevelt willow		158	Threadleaf groundsel
120	Running live oak		160	Twinleaf senna, twoleaf senna
122	Sacahuista		162	Twisted acacia, huisachillo
124	Saltcedar, tamarisk		164	Upright prairie coneflower
126	Sand sagebrush		166	Western bitterweed
128	Sand shinnery oak, Havard shin oak		168	Western honey mesquite
130	Seepwillow		170	Western horse nettle, tread-salve
132	Sericea lespedeza		172	Western ragweed
134	Silverleaf nightshade		174	Whitebrush, beebush
136	Skunkbush sumac, fragrant sumac		176	White-thorn acacia, mescat acacia
138	Smartweed		178	Wild carrot
140	Spiny aster, wolfweed, Mexican devil-weed		180	Willow
142	Spiny hackberry, granjeno		182	Willow baccharis
144	Sulfaweed, broadleaf sumpweed		184	Winged elm
146	Sweetgum		186	Woolly locoweed
148	Tarbush		188	Yankeeweed, rosinweed
150	Tasajillo, turkey pear		190	Yaupon
152	Texas bullnettle, mala mujer		192	Yucca
			194	**Range Plants by Region**
			200	**Index**

Preface

About 111 million acres in Texas—more than 70 percent of the state's land surface—is considered native rangeland or permanent pastures. These lands provide forage for livestock, habitat for wildlife, watersheds for streams and reservoirs, and open space for recreation.

More than 87 percent of Texas native rangelands are infested with unwanted weed and brush species at varying densities. When these plant populations are too dense, they compete with desirable forage plants for soil, nutrients and rainfall and reduce the value of the land for hunting and recreation. They also make it difficult to handle livestock or move over the land in these areas.

Rangeland sustainability can be maintained or enhanced by using brush and weed management systems and practices that are economically viable, environmentally sound and socially acceptable. Thousands of these rangeland acres are managed each year, and AgriLife Extension range specialists are often asked for advice on control strategies for specific plant species.

Since the early 1990s, Texas land managers, county Extension agents and other agency personnel have relied on Extension publication B-1466, *Chemical Weed and Brush Control Suggestions for Rangeland*, as the au-

Preface (continued)

thority for individual species control recommendations using herbicides. Originally developed by Tommy Welch, former Extension Range Specialist, the publication is updated annually by a team of specialists within the Department of Ecosystem Science and Management at Texas A&M University. It has become one of the most requested Extension publications.

Successful brush or weed management plans begin with proper plant identification. *Brush and Weeds of Texas Rangelands* was developed to aid in identifying the plant species covered in B-1466. This guide provides a description, identifying characteristics, the geographic distribution and a habitat description of each species, along with color photographs of the plant and its distinguishing features. This book does not give specific management recommendations; for those, please see B-1466. This publication and others are available on the Internet at **http://agrilifebookstore.org**.

— *Charles R. Hart*

Vegetational Areas of Texas

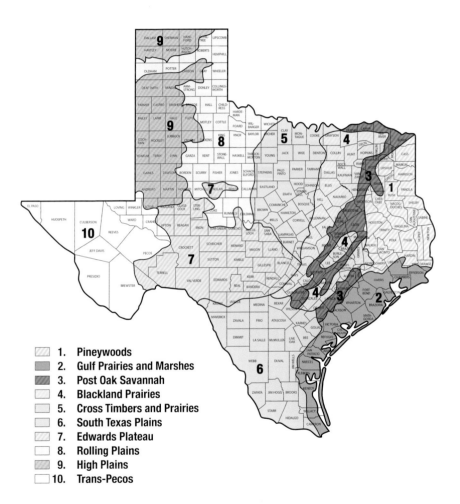

1. Pineywoods
2. Gulf Prairies and Marshes
3. Post Oak Savannah
4. Blackland Prairies
5. Cross Timbers and Prairies
6. South Texas Plains
7. Edwards Plateau
8. Rolling Plains
9. High Plains
10. Trans-Pecos

Note: The colored sections of the maps on the plant description pages show the areas where each plant grows.

Plant Descriptions

African rue

Peganum harmala L.

Family: Zygophyllaceae (Caltrop family)

African rue is an introduced, deep-rooted, perennial plant in the Caltrop family. The plant is a succulent, bright green herb growing from a woody base. It is bushy, many branched and about 1 foot tall when fully grown.

The leaves of African rue are hairless, divided into narrow segments and located alternately along the stems. The flowers consist of five white petals and are present from April to September. Segmented seed pods develop from May to October.

The forage value of African rue is poor for livestock and wildlife, and the plant is extremely unpalatable. It is also poisonous to livestock and is consumed only when animals are starving

Distribution: Regions 7, 8 and 10

Habitat: African rue is found in a variety of soil types but thrives in disturbed areas, roadsides and abandoned cropland.

Annual broomweed, common broomweed

Amphiachyris dracunculoides (DC.) Nutt.
Family: Asteraceae (Sunflower family)

Annual broomweed is a warm-season native plant in the Sunflower family. It is an annual that has a single stem growing 15 to 30 inches tall. The plant branches near the top to form a uniform crown with small yellow flowers in September through November. This growth arrangement distinguishes annual broomweed from perennial broomweed.

The first leaves usually are lance-shaped and form along the main stem, shedding when the crown begins to develop. Older leaves are fine and located alternately along the branch stems.

Annual broomweed provides poor grazing for wildlife and livestock.

Distribution: Regions 1–10

Habitat: The habitat is broad, ranging from dry native ranges to disturbed sites. It is found most typically on heavy clay soils.

Flower

Whole plant

Seedling

Population

Annual croton: One-seed croton

Croton monanthogynus Michx.
Family: Euphorbiaceae (Spurge family)

Of the 20 croton species occurring in Texas, nine are annuals. Annual crotons range in size from small to large and have star-shaped hairs or shield-shaped scales on at least some plant parts.

One-seed croton is a native, warm-season annual that is also commonly named prairie-tea. It can reach several inches to 18 inches tall, depending on moisture conditions.

The plant has a taproot, many wide branches, and stems that are usually a peachy-pink-orange color. Although most members of this plant family have a three-seeded capsule or fruit, the one-seed croton has a capsule with only one seed.

One-seed croton produces many seeds for seed-eating birds but is seldom grazed by livestock.

Distribution: Regions 1–8 and 10

Habitat: These annual plants usually grow most abundantly where there is soil disturbance, lack of soil cover or overgrazing. One-seed croton occurs in abundance on calcareous soils.

Whole plant

Flower

Leaf

Annual croton: Texas croton

Croton texensis (Klotzch) Muell. Arg.
Family: Euphorbiaceae (Spurge family)

Like the other native, annual crotons, Texas croton has an aromatic smell when the leaves are crushed. It varies from 1 foot to 4 feet tall, depending on moisture conditions.

The leaves are grayish to yellowish green and may be lighter on top and darker beneath. They are usually entire or without lobes or teeth and are located alternately along the stems. Each leaf is attached to the stem by a small stalk called a petiole.

The flowers are arranged in spikes at the ends of the stems. The fruit of Texas croton is a capsule divided into three segments supporting three individual seeds.

Texas croton produces a seed crop that is very valuable to dove, quail and other seed-eating birds but has low value for livestock grazing.

Distribution: Regions 2–10

Habitat: Texas croton grows on calcareous soils, sandy loam soils and loose sands. It can occur in great abundance and is generally associated with soil disturbance, lack of soil cover or overgrazing.

Fruit

Whole plant

Flower

Leaf

Annual croton: Woolly croton

Croton capitatus Michx.

Family: Euphorbiaceae (Spurge family)

Twenty croton species occur in Texas. Of these, nine species are annuals. Woolly croton is a native, warm-season annual with star-shaped hairs on the surfaces of the leaves and stems.

Each leaf is attached to the stem by a stalk, called a petiole. The leaves have no lobes, are usually entire and are located alternately along the stem. They have an aromatic smell when crushed.

The flowers are arranged in spikes at the ends of the stems. The fruit is a capsule that has three segments supporting three individual seeds.

Woolly croton produces seeds that are very valuable to dove, quail and other seed-eating birds but are of low value for livestock grazing.

Distribution: Regions 1–10

Habitat: When these plants are abundant, they are generally associated with soil disturbance, lack of soil cover or overgrazing.

Whole plant

Leaf

Flower

Population

Ashe juniper, blueberry juniper

Juniperus ashei Buchholz
Family: Cupressaceae (Cypress family)

This evergreen tree is a perennial, cool- and warm-season native. It is commonly called cedar but is not a true cedar. Ashe juniper produces limbs from the main trunk that are generally straight and are used commercially as fence posts.

The slender leaves are sharp pointed, each with a gland on the back that can be round or elliptical, pink or greenish. On long leaves, these glands are elongated and tapering. From seedling to 2 years old, Ashe juniper bears leaves that are awl-shaped or needlelike.

The blue-black berry has a waxy, white bloom. The dark gray or brown bark breaks into long, pliable strips.

Ashe juniper provides fair grazing for wildlife and poor grazing for livestock.

Distribution: Regions 5–10

Habitat: An aggressive invader, ashe juniper is found mostly on limestone and shallow, rocky soils in Texas. It is most commonly found in the Central Texas region.

Whole plant

Fruit

Stem

Seedling

Berlandier lobelia

Lobelia berlandieri A. DC.
Family: Campanulaceae (Bluebell family)

Berlandier lobelia is a cool-season annual of the Bluebell family that can reach up to 20 inches tall; however, most plants are less than 12 inches tall.

Most of its leaves originate from the base of the stem and are hairless, oval and up to 2 inches long on short stems. The leaves on the stalks are lance-shaped and much smaller. Berlandier lobelia has from one to 20 thin, ascending branches, each bearing a loose stalk of small, bright, purplish-blue flowers with white "eyes."

The plant is poisonous to both livestock and wildlife.

Distribution: Regions 2 and 6

Habitat: In Texas, these plants are confined to the Rio Grande Plains, but they extend south into Mexico. They are abundant on disturbed rocky, sandy or clay soils in years with adequate fall and winter rainfall.

Whole plant

Flower

Leaf

Bigelow shinoak, scalybark oak, white shinoak

Quercus durandii Buckl. var. *breviloba* (Torr.) Palmer

Family: Fagaceae (Beech family)

Bigelow shinoak is commonly referred to as shin oak or scrub oak because of its multi-stemmed growth forming dense thickets about 10 to 12 feet tall. It is a member of the Beech family and one of the deciduous oaks.

The leaves are located alternately along the stem and vary greatly in shape and size. They have few to many lobes and are usually broadest above the middle of the leaf.

The bark is gray and rough, flaking off in strips. Young twigs tend to be greenish gray or brown. Acorns are usually solitary or in clusters of two or three on short stems.

The forage value of this plant is poor for cattle, but it is grazed occasionally by goats and deer.

Distribution: Region 7

Habitat: Bigelow shinoak is found on dry, limestone hills of Central Texas.

Fruit

Whole plant

Stem

Leaf

Bitter sneezeweed, basin sneezeweed

Helenium amarum (Raf.) H. Rock var. *amarum*
Helenium amarum (Raf.) H. Rock var. *badium* (Gray ex S. Wats.) Waterfall

Family: Asteraceae (Sunflower family)

Bitter sneezeweed and basin sneezeweed are erect, upper-branching native annuals that reach 10 to 20 inches tall. The entire plant of both varieties has a strong odor and is bitter to the taste.

The leaves are narrow and located alternately on the stem. In some years, the lower leaves are lost, new growth occurs up the stalk, and new flowers may appear in the fall.

A member of the Sunflower family, the plant has showy flowers that are noticeable in the late spring or summer and are located at the end of each branch. Two varieties of this plant are identical except for the flower color: one is all yellow, the other is yellow with a red-brown center. Each bloom has about eight cleft ray flowers (petals) with three lobes, often bending downward at maturity.

The plant is toxic to grazing livestock but is rarely consumed in toxic amounts.

Distribution: Regions 1–10

Habitat: The yellow variety is widespread in disturbed, sandy or loamy soil in the eastern to central part of the state; the dark-centered variety is often found in calcareous soil in more central and western areas of the state.

Bitter s. population

Bitter s. whole plant

Bitter s. flower

Basin s. flower

Basin s. whole plant

Basin s. population

Blackberry, dewberry

Rubus spp. L.
Family: Rosaceae

Blackberries are native perennials that are sometimes referred to as dewberries. Ten species of blackberry are listed for Texas. For the most part, it is an upright, thicket-forming shrub that is prickly and can grow to several yards tall.

The plant can reproduce by seed and from roots, as well as by daughter plants when the end of a stem reaches the soil. The stems are usually green, purplish or red and covered with prickles.

The leaves are located alternately along the stems; each has five leaflets arranged in a starlike shape, with each leaflet having small teeth around the edges.

The flowers are about ¾ to 1 inch in diameter, with five white petals and five green sepals. They bloom in late spring and at the beginning of the summer. The fruit is a berry that when ripe is black and about ⅓ to 1 inch in diameter.

Blackberries are used by wildlife but have little to no value for livestock.

Distribution: Regions 1–5, 7 and 8

Habitat: This plant prefers moist soils and grows mostly in wooded areas of East and Central Texas.

Blackbrush

Acacia rigidula Benth.

Family: Fabaceae (Legume family)

Blackbrush is a perennial, warm-season native. This shrubby tree grows to 12 feet tall and produces limbs from a central trunk. The zigzag branches have short, straight thorns in pairs.

The leaves are twice compound, with both divisions having one to eight small leaflets, typical of the legume family. The flowers are white. The 2- to 4-inch-long seed pods are narrow, curved and flat, have divisions between the seeds, and are reddish brown when ripe.

This tree provides fair grazing for wildlife and poor grazing for livestock.

Distribution: Regions 2, 6, 7 and 10

Habitat: Blackbrush is common on South Texas brushland. It is most often found on sandy or calcareous soils and on limestone caliche ridges and hills.

Whole plant

Spiny stem

Leaf

Mature stem

Fruit

Flower

Blackjack oak

Quercus marilandica Muenchh.
Family: Fagaceae (Beech family)

This large hardwood tree is a perennial, warm-season native with bark that is nearly black, very rough and arranged in ridges on the trunk. It is a member of the Beech family.

The leaves are scalloped with short, white hairs on top and brownish fuzz underneath. The leaves have three to five lobes, each with a short bristle on the tip.

Blackjack oak provides fair grazing for wildlife. For livestock, it is poisonous and provides poor grazing.

Distribution: Regions 1–4, 7 and 8

Habitat: Blackjack oak prefers slightly to very acidic sand, sandy loam and clay soils in Central and East Texas. In the western region of its boundaries, it may grow on gravelly clay soils. It is commonly associated with mesquite and juniper or other species of oak trees.

Whole plant

Leaf

Stem

Bois d'arc, osage orange

Maclura pomifera (Raf.) Schn.

Family: Moraceae (Mulberry family)

Bois d'arc, also commonly named osage orange, is a small to medium-sized tree in the Mulberry family growing to 60 feet tall. It is a perennial, cool-season native with white, milky sap.

The deciduous leaves are simple, range from 2 to 4 inches long, and are located alternately along the stem. Young leaves can have a covering of fine hair. The older leaves are hairless and lustrous.

Each twig or branch is armed with stout spines on the angle between the upper side of the twig and the supporting branch. The flowers are very small, greenish and arranged singly along an elongated, unbranched axis.

The noticeable fruit is a syncarp, which is a fruit consisting of many individual small fruits or drupes, such as a blackberry or pineapple. The fruit is globe-shaped or round, yellowish green and 4 to 5 inches in diameter. Its juice is a milky acid.

Bois d'arc was once planted largely for windbreaks or hedgerows. Squirrels feed on the fruit, and whitetail deer and goats browse the leaves.

Distribution: Regions 1–5, 7, 8 and 10

Habitat: Bois d'arc grows in deep, moist to semi-moist soils, favoring rich clay soils.

Fruit

Mature stem

Whole plant

Leaf

Spiny stem

Broom snakeweed, perennial broomweed

Gutierrezia sarothrae, (Pursh.) Britt. & Rusby, *G. microcephala* (DC.) Gray

Family: Asteraceae (Sunflower family)

Perennial broomweed or broom snakeweed is a short-lived, perennial half-shrub in the Sunflower family. It ranges from 6 inches to about 2 feet tall.

The shrub has many unbranched, erect stems originating from a woody base that die back when the plant goes dormant. The leaves are narrow and threadlike. From June to October, small yellow flowers are clustered at the branch tips.

The plant's forage value for livestock and wildlife is poor, and perennial broomweed is poisonous to livestock, causing abortions when consumed during late pregnancy. It can cause death if consumed as more than 20 percent of the diet.

Distribution: Regions 2–10

Habitat: Perennial broomweed is widespread on dry ranges and deserts from California to Texas, south to Mexico and north to Idaho. Extreme infestations reduce forage production but may not indicate overgrazed ranges because broomweed populations fluctuate naturally. However, overgrazing does accelerate the plant's growth and propagation.

29

Buffalo-bur

Solanum rostratum Dun.

Family: Solanaceae (Nightshade family)

A prickly, annual, warm-season plant of the Nightshade family, buffalo-bur typically can grow to 2 feet tall. This plant is considered a weed nearly everywhere it grows.

The leaves, which vary in shape and size, are irregularly rounded and deeply lobed and have spiny veins. The stems are profusely thorned. The yellow flowers appear from May to October, and the fruit is enclosed by a prickly bur.

Buffalo-bur provides only fair grazing for wildlife and is poisonous to livestock. Because of its spiny growth form, it is rarely consumed.

Distribution: Regions 1–10

Habitat: Buffalo-bur is common in old fields, roadsides, overgrazed pastures and disturbed areas and near water tanks throughout Texas. A native of the Great Plains, it is found from North Dakota to Texas and westward and south into Mexico.

Burrobrush

Hymenoclea monogyra T. & G. *ex* Gray.
Family: Asteraceae (Sunflower family)

Burrobrush is a native perennial that is also called cheeseweed or cheesebrush because when crushed, the leaves give off an obviously cheeselike odor. This upright shrub usually grows to between 2 and 8 feet tall.

Burrobrush leaves are located alternately along the stem and are generally threadlike, dark green and about 1 to 3 inches long. The thin stems arise from a single base and often arch out and bend toward the ground.

The plant produces clusters of white flowers that bloom in the fall; female and male flowers can occur in the same cluster. The seeds are formed within a winged bur.

Livestock typically avoid burrobrush.

Distribution: Regions 7 and 10

Habitat: Burrobrush is found on dry and well-drained sites with alluvial soils in the Trans-Pecos and Edwards Plateau areas.

Population

Whole plant

Leaf

Flower

Camphorweed

Heterotheca subaxillaris (Lam.) Britt. & Rusby var. *latifolia* (Buckl.) Gandhi & Thomas

Family: Asteraceae (Sunflower family)

Because its name is used for different plants across the United States, camphorweed must be checked by scientific name for proper identification. Also known as *H. latifolia* by some authors, it is generally recognized by the strong camphor-like scent when the leaves are crushed.

Camphorweed is an annual, warm-season native that generally emerges from the ground as a single stem, then branches several inches above the ground.

Its stems and leaves are covered in spreading white hairs. It has showy, bright yellow flowers with hairy leaves clasping the spindly branches. It flowers from May through July, depending on moisture.

The forage value of camphorweed is fair for most grazing animals.

Distribution: Regions 1–10

Habitat: Camphorweed can grow profusely in disturbed sites and a variety of soil types, preferring sandy soils.

Carolina horse nettle

Solanum carolinense L.
Family: Solanaceae (Nightshade family)

Carolina horse nettle is a coarse, branching, warm-season perennial in the Nightshade family. It grows 1 to 3 feet high.

Carolina horse nettle has large spines on the stems and leaves. Each of its mostly oval leaves has several large teeth or shallow lobes on both sides. On the underside are microscopic, star-shaped hairs.

The clustered flowers are pale violet to white and give rise to spherical fruit. The fruit is about ½ inch in diameter; it is green with light green vertical bands until maturity, when it becomes uniformly yellow.

Because of its many spines, the forage value of Carolina horse nettle is poor for wildlife and livestock. The fruit is poisonous to livestock.

Distribution: Regions 1–4 and 7

Habitat: This plant grows across the eastern part of Texas and the entire eastern half of the United States. It is found mostly in sandy soils in fields, open woodlands and waste places.

Fruit

Flower

Whole plant

Catclaw acacia

Acacia greggii Gray.
Family: Fabaceae (Legume family)

A thorny, thicket-forming, native shrub or small tree in the Legume family, this plant occasionally grows to 30 feet tall, with trunks of up to 12 inches in diameter. However, these trees are usually much smaller.

The leaves are doubly pinnate, with small leaflets on one side and then on the other at different levels on the stem. The fragrant, pale white flowers generally occur from April to October. They develop into curling, contorted fruit pods that persist from July through winter.

Catclaw acacias have many alternating thorns pointing upward that make dense thickets impenetrable.

Its forage value is fair for wildlife and goats and poor for cattle and sheep.

Distribution: Regions 2, 5–8 and 10

Habitat: Catclaw acacias grow on dry, gravelly mesas with shallow caliche and in arroyos and deep, alkali sand. The plant is also extremely drought tolerant.

Whole plant

Flower

Stem

Leaf

Fruit

Catclaw mimosa, fragrant mimosa

Mimosa biuncifera Benth.
Mimosa borealis Gray
Family: Fabaceae (Legume family)

These spiny, thicket-forming, native shrubs in the Legume family grow to 8 feet tall. The spines of catclaw mimosa are stout, upward pointing, flat at the base and generally paired at each node. The stems are typically flexible, angled and alternating direction at each node. Fragrant mimosa differs, having single thorns and straight stems.

The long, crowded leaves are made up of many opposite leaflets that are hairless. The leaves originate at each node on a short pedicel, or small stalk. Pink to whitish rounded flowers occur in the spring and are very fragrant. By September, the fruits mature into linear, curved or straight bean pods.

The forage value of these shrubs is poor for livestock and fair for wildlife.

Distribution: Regions 5 and 7–10

Habitat: Catclaw mimosa and fragrant mimosa are found mainly on dry hills and mesas in Central and West Texas.

Catclaw m. stem

Fragrant m. whole plant

Catclaw m. leaf

Fragrant m. flower

Ceniza, purplesage, Texas silverleaf

Leucophyllum frutescens (Berl.) I.M. Johnst.
Family: Scrophulariaceae (Foxglove family)

Ceniza is a short, unarmed shrub growing to 10 feet tall. A member of the Foxglove family, ceniza is a colorful plant that usually stands out on native range, especially when flowering. It is used extensively as an ornamental.

The leaves are hairy, ashy gray, simple and either cluster along the stems or are located alternately along the stems. The flowers are showy, pale violet to purple or pink and have external and internal hairs. On old trunks, the bark is grayish black and rough with small scales.

The forage value for ceniza is fair for livestock and wildlife.

Distribution: Regions 2, 6, 7 and 10

Habitat: In Texas, ceniza is common on rocky limestone or caliche hills, and in arroyos and chaparral.

Leaf

Population

Whole plant

Flower

43

Chinese tallow tree

Sapium sebiferum (L.) Roxb.
Family: Euphorbiaceae (Spurge family)

Chinese tallow tree is a fast-growing weedy tree with milky sap. It grows to 30 feet tall and often spreads by root sprouts. Its slender limbs and branches droop and are easily broken.

The leaves of Chinese tallow trees are hairless and located alternately along the stem. The leaves have smooth margins and diamond-shaped blades that are shorter than the petioles (leaf stalks). They turn bright yellow, orange or red in the fall.

The flowers have no petals and grow in 2- to 6-inch drooping spikes at the end of each branch. The fruit has three cells, the walls of which fall readily at maturity. The three chalky white seeds may remain attached to the cells through the winter. These nutlike seeds have a hard coat covered by tallow that blackens with weathering.

The forage value of Chinese tallow trees is poor for livestock and fair for wildlife.

Distribution: Regions 1 and 2

Habitat: Introduced from Asia, Chinese tallow is planted widely as an ornamental. Birds disperse the seeds, and it has escaped in the southeastern part of Texas, where it can be a significant invading woody species.

Fruit

Whole plant

Leaf

Flower

Cholla

Opuntia imbricata (Haw.) DC.
Family: Cactaceae (Cactus family)

Cholla is a large, upright cactus reaching a height of 10 to 13 feet. This plant is often called walking stick cholla. It is very spiny with barbed white to green spines, about ¾ to 1¼ inches long.

The flowers are yellow to pink, 1½ to 2¾ inches long and located on the ends of the stems. The fruit is yellow, dry and produces seeds a little over 1/10 inch long.

Cholla is distributed from Texas northward to Oklahoma and Kansas and westward to New Mexico.

The fruit provides fair forage for wildlife but is rarely consumed by grazing livestock.

Distribution: Regions 7–10

Habitat: Cholla occurs on clay and clay loam soils or foothills in West Texas, especially the High Plains and Trans-Pecos regions.

Population

Flower

Whole plant

Stem

Fruit

Christ thorn, Jerusalem thorn

Paliurus spina-christi Mill.
Family: Rhamnaceae (Buckthorn family)

Christ thorn is a shrubby tree from Asia that has become established in certain riparian (associated with the banks of a river or stream) areas in the central Edwards Plateau. It is an introduced, cool-season perennial.

This multi-stemmed shrub in the Buckthorn family can grow to 15 feet tall. Christ thorn can form dense, impenetrable thickets.

The plant has yellow fruit shaped like a hat, and paired thorns, one straight, one curved.

Christ thorn is not readily browsed by livestock or deer.

Distribution: Mainly in Region 7

Habitat: Christ thorn grows along riparian areas and low areas that receive extra moisture.

Flower

Whole plant

Leaf

Cocklebur

Xanthium strumarium L.
Family: Asteraceae (Sunflower family)

Cocklebur is an introduced annual plant in the Sunflower family. It is a coarse, rough weed commonly found across Texas. This plant spreads rapidly around tanks and down draws when moisture is adequate for germination.

The leaves are toothed or lobed and are located alternately along the stem. Separate male and female flowers grow on the same plant, although both are inconspicuous. The male flowers occur in dense clusters on the ends of the stems; female flowers occur in the leaf axils.

Cocklebur fruits are conspicuous and covered with many spines. The fruit has two compartments, each containing a seed.

The plant's forage value for wildlife and livestock is poor, and cocklebur in the seedling stage is poisonous to livestock.

Distribution: Regions 1–10

Habitat: Cocklebur is found throughout most of the United States. In dry regions, it is most common around water holes, playas, arroyos and disturbed areas.

Common goldenweed, Drummonds goldenweed

Iscoma coronopifolia (Gray) Greene,
I. drummondii (T. & G.) Greene

Family: Asteraceae (Sunflower family)

Common goldenweed and Drummonds goldenweed are native, warm-season, perennial sub-shrubs with woody taproots.

Common goldenweed usually has leaves that are pinnately (arranged on a common axis) lobed with three or four deep lobes. It can also have clusters of small leaves in the leaf axils. The leaves of Drummonds goldenweed are seldom lobed and may have only a few prominent teeth on the leaf margins.

Leaflike bracts occur below the flower heads of both species. The bracts of common goldenweed are usually about ¼ inch long; those of Drummonds goldenweed are commonly ¼ to ½ inch long.

Each species has a round-topped mass of flowering heads. The flowering heads are yellow, and the plants have no ray or showy flowers.

Distribution: Regions 2 (Drummonds goldenweed) and 6 (both species)

Habitat: Common goldenweed is distributed throughout the western part of the South Texas Plains on dry, open, calcareous soils. Drummonds goldenweed occurs on the coastal half of South Texas on various soil types; it is considered a more aggressive weedy species.

Whole plant

Flower

Stem

Leaf

Common persimmon, eastern persimmon

Diospyros virginiana L.
Family: Ebenaceae (Persimmon family)

Common persimmon is a shrub to small tree. A native, cool-season perennial, it is also known as eastern persimmon.

This plant has a characteristic leaf shape, bud scars and new buds. Its fruit has four to eight seeds that are very hard and enable this plant to invade pasture land. As it matures, the fruit is at first green, turning yellow to orange and finally purple to black when fully ripened.

Common persimmon provides fair forage for wildlife.

Distribution: Regions 1–5, 7 and 8

Habitat: Common persimmon grows on most types of soils from sand to shale and in muddy bottomlands in the eastern and central parts of Texas.

Common sunflower, annual sunflower

Helianthus annuus L.
Family: Asteraceae (Sunflower family)

This tall, showy plant grows from 2 to 8 feet tall. Its stem has short, prickly hairs and may sometimes be purple or dark colored. The leaves are covered in coarse, rough hairs and are located alternately along the stem. The leaf blades can grow to 10 inches long.

The flower heads are terminal (growing at the end of a branch or stem) and large—up to 4 inches across. The ray flowers are yellow and the disk flowers brown. The plant blooms from May through October.

The forage value of common sunflower for livestock is poor, but it provides excellent feed for birds.

Distribution: Regions 1–10

Habitat: This plant grows in many soil types and moisture conditions. It is prominent on disturbed and low-lying areas but can also persist on dry soils.

Flower

Seedling

Population

Whole plant

Leaf

Creosotebush, greasewood

Larrea tridentata (DC.) Cov.
Family: Zygophyllaceae (Caltrop family)

An evergreen, aromatic shrub in the Caltrop family, creosotebush generally grows from 3 to 6 feet high and sometimes to 11 feet. It is extremely shallow rooted and drought tolerant.

When mature, the stems are rough and dark gray to black; young twigs are brown and flexible with large, dark nodes that give the plant a jointed appearance.

The leaves of creosotebush are dark green, pointed at the tips and situated in pairs across from each other on the stem. If rainfall is adequate, yellow flowers occur from spring through summer.

The shrub provides poor forage for livestock and wildlife.

Distribution: Regions 6, 7 and 10

Habitat: Creosotebush is commonly found in shallow soils with underlying hardpan (a layer of hard soil or clay) in the Trans-Pecos region of Texas.

Population

Flower

Leaf

Stem

Whole plant

Curlycup gumweed

Grindelia squarrosa (Pursh) Dun.
Family: Asteraceae (Sunflower family)

A member of the Sunflower family, curlycup gumweed is a weedy, warm-season perennial. The plant may be branched or unbranched, and it can vary from 8 to 35 inches tall.

Curlycup gumweed starts growing in late spring, begins to flower in July or August, and dries up in late summer. The leaves are hairless, shiny and heavily toothed along the margins. They tend to clasp the stem. The flowers are yellow and very sticky below the disk flowers.

The plant has been suspected of accumulating selenium. It has poor grazing value for wildlife and, when young, fair value for livestock.

Distribution: Regions 1, 3–5 and 7–10

Habitat: This plant grows mainly in waste areas and disturbed areas as a weedy perennial.

Dog cactus, dog cholla, clavellina

Opuntia schottii Engelm.
Family: Cactaceae (Cactus family)

Also called dog pear or clavellina, this plant is a mat-forming cactus, with the clumps reaching 4 to 6 inches tall and 3 to 9 feet in diameter. Dog cactus is an extreme nuisance, as its stems break off when stepped on or brushed against by livestock, wildlife or people.

The joints of this plant are ¾ to 2½ inches long, with many 1- to 2-inch-long spines that are brownish or gray tinged with pink or red. The spines are sheathed (encased with a protective covering) only at the tips.

The flowers and fruit are yellow; the fruit is fleshy and expanded upward.

The forage value of dog cactus is poor for livestock and wildlife.

Distribution: Regions 5–7 and 10

Habitat: This plant grows in sandy to sandy loam soils at elevations of 900 to 4,500 feet in West Texas and New Mexico.

Eastern baccharis

Baccharis halimifolia L.
Family: Asteraceae (Sunflower family)

Eastern baccharis is a native, warm-season perennial of the Sunflower family. Its other common names are sea-myrtle and consumption-weed. This shrub can reach 6½ feet tall.

Eastern baccharis often occurs abundantly in open, sandy places. This plant can also be an invader of old farmland.

The leaves of the plant are simple and arranged alternately on the stem. The leaf edges are smooth or have a few remote, coarse, angled teeth. On the stems supporting the flowers (inflorescence), the leaves are smaller and narrower. Generally, the leaf blade can be from 1 to 3 inches long and ¼ to 2 inches wide.

The forage value is poor for livestock and wildlife.

Distribution: Regions 1–3

Habitat: Eastern baccharis grows in open, sandy areas and arroyos.

Flower

Whole plant

Leaf

Eastern red cedar

Juniperus virginiana L.
Family: Cupressaceae (Cypress family)

An evergreen tree, eastern red cedar is a member of the Cypress family. It is an aggressive invader that spreads mainly by seed.

Although variable, its shape tends to be like a Christmas tree. Eastern red cedar generally reaches 20 to 30 feet tall. Its bark is light reddish brown and separates into long strips.

The small, purplish flowers are deciduous, occurring from March to May. The fruits ripen from September to December. The berrylike fruits are pale blue, smooth and sweet to the taste.

The forage value of eastern red cedar is poor for livestock and fair for wildlife.

Distribution: Regions 1–4 and 7

Habitat: Eastern red cedar grows in all types of soils, from hilltops to swamps.

Stem

Fruit

Whole plant

Leaf

Elm

Ulmus spp. L.
Family: Ulmaceae (Elm family)

Four species of native elms are listed in Texas: winged, American, cedar and slippery elm. Depending on age and other conditions, elms can be small, shrubby trees or large trees reaching up to 90 feet tall. The bark usually has deep, vertical furrows.

The leaves grow on short petioles (leaf stalks). Along the margins, the leaves are heavily veined and sharply toothed; at the base they are unevenly shaped. The flowers are small and generally not noticeable. The seeds have wings for wind dispersal.

The leaves offer fair forage value for whitetail deer and goats.

Distribution: Regions 1–8 and 10

Habitat: Elms grow in high-moisture areas along creeks and streams and around ponds. Some species are found in drier areas such as fence lines and abandoned fields in East Texas. They grow in most soil types but prefer neutral to acidic sands and sandy loams. Elms are commonly found growing with hackberry.

Stem

Whole plant

Leaf

Mature stem

Flower

Flameleaf sumac

Rhus copallina L.
Family: Anacardiaceae (Sumac family)

Flameleaf sumac is a slender-branched shrub or small deciduous tree in the Sumac family. This species usually grows in small mottes or clusters, as the plant can spread by rhizomes (horizontal, usually underground stems that often send out roots and shoots from the nodes).

The leaves are long, narrow, compound and located alternately along the stem. The leaves turn a showy shade of red during the fall. Younger stems are covered in small hairs, but older stems become hairless.

In early summer, small, whitish flower clusters occur on the tips of the branches and mature into rounded, red fruits.

The forage value of flameleaf sumac is fair for goats and wildlife.

Distribution: Regions 1–5 and 7

Habitat: This plant grows mainly on rocky hills, woods and bottomlands in the eastern, central and southern parts of Texas.

Whole plant

Flower

Mature stem

Fruit

Leaf

Garboncillo, rattleweed

Astragalus wootonii Sheldon
Family: Fabaceae (Legume family)

Garboncillo is a much-branched annual in the Legume family. It has erect, hairy stems reaching 3 to 12 inches long.

The leaves are composed of nine to 19 leaflets, hairy beneath and smooth above. The flowers are pink or purplish to white, emerging from March to June.

The most conspicuous part of the plant is the fruit, a large, one-celled, inflated pod. When dry, the seeds in the pod break loose and rattle, giving it the name rattleweed.

This plant is poisonous to livestock and wildlife.

Distribution: Regions 9 and 10

Habitat: In Texas, garboncillo is generally restricted to clay soils in the Trans-Pecos region. It is also common in southern New Mexico, eastern Arizona and northern Mexico. Garboncillo is most abundant in valleys receiving runoff water from the surrounding hills, as well as in bar ditches, along trails and around earthen tanks.

Fruit

Whole plant

Flower

Leaf

Gray goldaster

Heterotheca canescens (DC.) Shinners
Family: Asteraceae (Sunflower family)

A common plant throughout most of Texas, gray goldaster is an aggressive perennial of the Sunflower family. It is a bushy, much-branched plant, generally reaching 4 to 18 inches tall.

The leaves are numerous and crowded along the plant stems. Because of the extremely dense hairs on its leaves, gray goldaster is difficult to control with herbicides.

The yellow flower heads occur at the terminals (ends) of each branch and are ⅜ to ⅝ inch in diameter, blooming from July through September.

The forage value of gray goldaster is poor for livestock and fair for wildlife.

Distribution: Regions 2 and 4–10

Habitat: Gray goldaster is found in sandy and gravelly prairies and rock outcrops. It is a common roadside wildflower that grows in colonies.

Leaf

Whole plant

Stem

Flower

Greenbriar, saw greenbriar

Smilax bona-nox L.
Family: Liliaceae (Lily family)

Greenbriar is a tough, woody, high-climbing vine in the Lily family. It spreads aggressively from long, slender rhizomes, which are horizontal, usually underground stems that often send out roots and shoots from the nodes.

Along the stems are stout, flattened prickles. The numerous tendrils are used for climbing. The leaves have short petioles (stems) and are hairless and bright green on both sides, with rounded to heart-shaped bases.

The flowers are greenish to bronze, and the berries are green when young and blue-black at maturity, each with two or three seeds.

When greenbriar is young and succulent, its forage value is fair for goats and wildlife.

Distribution: Regions 1–8

Habitat: Greenbriar is found trailing over trees, shrubs and fences and in rolling woodlands in Central to East Texas.

Stem

Whole plant

Leaf

Fruit

Guajillo

Acacia berlandieri Benth.
Family: Fabaceae (Legume family)

Guajillo is a nearly thornless shrub to small tree in the Legume family. The plant height varies greatly, but it can grow up to 15 feet tall.

Guajillo leaves are arranged like those of mesquite but are smaller. The sweet-scented, white to yellow flowers are clustered in dense groups. The plant produces a flattened, bean-type legume fruit that is four to six times longer than it is wide.

The forage value of guajillo is poor for livestock and wildlife except under drought conditions. However, it may be toxic when consumed as a major part of the diet.

Distribution: Regions 2, 6, 7 and 10

Habitat: Guajillo grows on a variety of soil types but is most prolific on ridges and shallow soils. Found mainly in the South Texas Plains and southwest Texas, it is less common in the southern Edwards Plateau and Trans-Pecos regions. It also grows extensively in northern Mexico.

Fruit

Population

Whole plant

Stem

Flower

Leaf

Hackberry

Celtis spp. L.
Family: Ulmaceae (Elm family)

Hackberry species occur throughout Texas; five species are trees and one species is shrublike. The two species most common across the state are *Celtis laevigata*, also called sugarberry or sugar hackberry, and *C. reticulate*, also known as netleaf hackberry or western hackberry.

The trees have strong tap roots and many shallow, spreading roots. The bark is mostly smooth and gray, with small bumps or warts on the older stems. The wood has a characteristic yellowish white color.

The leaves of hackberry have a rough texture, like sandpaper. The leaf underside has large, netlike veins. Although not noticeable, the flowers occur in early spring and develop into rounded, succulent, reddish brown fruits (drupes) that persists on the tree throughout the winter.

The forage value is fair for wildlife and poor for livestock.

Distribution: Regions 1–10

Habitat: Hackberry grows in rocky draws and arroyos and other low areas receiving adequate moisture.

Fruit

Leaf

Whole plant

Stem

Hercules club, toothache tree

Zanthoxylum clava-herculis L.
Family: Rutaceae (Citrus family)

Hercules club is a small to medium-sized tree or shrub in the Citrus family. It is a native, cool-season perennial. The bark of the trunk is light gray and thinly covered with conspicuous, corky, cone-like tubercles (wartlike outgrowths). The twigs are brown to gray and have simple spines.

Hercules club leaves are located alternately along the stems and are pinnately (arranged on a common axis) compound, with spines on the supporting leaf stem. There are usually five to 19 leaflets. The leaves are lustrous, have serrated edges and may be somewhat hairy underneath.

The flowers are greenish white and appear in April and May. The mature fruit is a shiny black seedpod with one seed; the seeds occur in clusters.

Hercules club provides an excellent source of seeds and fruit for birds but has low value for grazing.

Distribution: Regions 1–4 and 7

Habitat: This tree commonly grows along fence rows and hedge rows on sandy and clay soils.

Leaf

Whole plant

Fruit

Stem

Thorny stem

Honey locust

Gleditsia triacanthos L.
Family: Fabaceae (Legume family)

Honey locust is a medium-sized to large tree that can reach 100 feet tall. It is a member of the Legume family and can grow in dense, impenetrable thickets.

On older trees, the bark is grayish brown to black, with clusters of thorns and deep cracks that separate into scaly ridges. The trunk and branches of honey locust are densely thorny. The thorns can sometimes be as long as 12 inches and are three pronged.

The leaves are deciduous and located alternately on the stems. They occur in a typical legume fashion, being twice pinnate. Showy flowers appear in May and June, and large bean pods up to 1½ feet long ripen in September and October.

The plant is considered poor forage for livestock and fair for wildlife.

Distribution: Regions 1–7

Habitat: Honey locust grows in moist, fertile soils across the eastern and central parts of Texas.

Thorny stem

Stem

Leaf

Fruit

Whole plant

Honey mesquite

Prosopis glandulosa Torr. var. *glandulosa*
Family: Fabaceae (Legume family)

Honey mesquite is a small to medium-height tree or shrub. It is thorny and has either one stem or branches near the ground. A member of the Legume family, it is the most common species of mesquite in Texas.

The leaves of honey mesquite are deciduous and located alternately along the stems. The fruits are loosely clustered pods (beans) reaching 8 to 10 inches long. They may be abnormally abundant in drought years.

The beans can provide fair forage for livestock and wildlife but can be toxic to livestock if consumed as a high percentage of the diet.

Distribution: Regions 1–10

Habitat: Generally found throughout Texas, honey mesquite is common on dry ranges across the state. It is found from California to Kansas, Texas and Mexico.

Leaf

Thorny stem

Mature stem

Whole plant

Flower

Fruit

Horehound

Marubium vulgare L.

Family: Lamiaceae (Mint family)

An erect perennial in the Mint family, horehound grows to 2½ feet tall. It is a native of Europe and considered invasive and generally unwanted.

The stems are woolly white, have the characteristic four edges of the mint family and are somewhat woody at the base.

The leaves appear wrinkled, are coarsely toothed along the margins and are arranged in opposite pairs on the stem. Small, white flowers and bur-like fruits develop in clusters in the leaf axils. The clusters of fruits are a severe nuisance in sheep wool.

The plant has fair value for wildlife and poor value for livestock.

Distribution: Regions 1–10

Habitat: Horehound is found along roadsides, dry waste areas and disturbed areas in most soil types across the state.

Flower

Whole plant

Stem

Leaf

Huisache

Acacia smallii Isely
Family: Fabaceae (Legume family)

Huisache is a small tree in the Legume family. It is shaped like an upside-down cone. This shrub to small tree is a native, warm-season perennial that is commonly named sweet acacia. The stems, which can reach 15 feet tall, have many spines that are paired, straight, pale and pinlike.

The leaflets are gray-green and twice compounded with eight to 16 divisions, each having 10 to 20 pairs of small, sensitive leaflets.

The flowers are produced on a fragrant, yellow, fluffy ball with many clusters of yellow stamens. Huisache fruit are black and tapered at each end. The seedpods are cylindrical and 1½ to 3 inches long; when mature, they turn dark brown or black.

Sometimes confused with twisted acacia, huisache can be distinguished by its spreading growth form, the shorter and broader legume, and the position and/or presence of a gland on the petiole, or leaf stem. In huisache, the petiolar gland is either absent or located near the middle of the leaf petiole.

Huisache provides poor grazing for wildlife and livestock.

Distribution: Regions 2–4, 6 and 7

Habitat: This tree grows in a variety of soils, most often on deep, poorly drained, sandy or clay lowlands.

Flower

Leaf

Whole plant

Stem

Whole plant, flowering

Lime pricklyash, colima

Zanthoxylum fagara (L.) Sarg.
Family: Rutaceae (Citrus family)

Lime pricklyash is a native, cool-season perennial. It is an evergreen shrub or small tree growing to 25 feet tall. The plant is very prickly and appears as a rounded shrub or a member of a brush motte, which is a grove or clump of trees in an open area.

The leaves are located alternately along the stems and are compound, with five to 13 leaflets arranged on a common axis and attached to a winged main stem. The leaflets are oval, from 1 to 2½ inches long and commonly have a toothed margin. When crushed, they smell like citrus.

The flowers are yellowish green and appear in late winter to spring. The fruit is a one-seeded pod that turns red or brownish red when ripe.

The leaves of Lime pricklyash are valuable browse for whitetail deer, and the seeds are eaten by seed-eating birds such as quail. The shrubs are also a favorite nesting site for several species of passerine (perching) birds.

Distribution: Regions 2 and 6

Habitat: Lime pricklyash grows in the brush and chaparral of the Rio Grande Plains and the coastal prairie on clay loam or sandy clay loam soils that are fairly well drained.

Leaf

Whole plant

Stem

Lotebush, blue brush, gumdrop tree

Zizyphus obtusifolia (T. & G.) Gray var. *obtusifolia*
Family: Rhamnaceae (Buckthorn family)

This shrub is a native, cool-season perennial in the Buckthorn family. Commonly named lotebush, gumdrop tree and clepe, it is a rigid, intricately branched shrub that grows to 6 feet tall and is greenish gray to blue-gray.

Lotebush has thorn-tipped branches of various lengths, as well as many sharp, straight thorns along the stems.

Lotebush leaves are often grayish green and vary greatly in shape from oval to oblong. They are ½ to 1½ inches long and either entire or shallow toothed. The flowers are inconspicuous. The fruit is black, mealy and stonelike, about the size of a pea.

This shrub provides good grazing for wildlife and poor grazing for livestock. It is best known for providing good cover for wildlife, especially quail, and a nursery area or protection for young plants that are easily grazed out of a pasture.

Distribution: Regions 2–10

Habitat: Lotebush is widespread across Texas on dry ranges in the central, southern and western parts of the state.

95

Whole plant

Population

Leaf

Stem

Macartney rose

Rosa bracteata Wendl.
Family: Rosaceae (Rose family)

Macartney rose is an introduced, warm-season perennial of the Rose family. It is an evergreen shrub that can grow to almost 10 feet tall. This plant is commonly considered an invasive species or pest. Historically, Macartney rose was planted on Texas landscapes as a living fence.

The stems have paired, very broad-based prickles. Each leaf is made up of five to nine tough, thick leaflets. The leaflets are lustrous above and a duller green beneath.

The flowers occur singly or in groups of one to three on short stalks. The flower petals are white. The fruit is round or spherical like that of other members of the Rose family.

Macartney rose has no grazing value for livestock or wildlife but may serve as escape cover for rodents and other small mammals.

Distribution: Regions 1–5 and 7

Habitat: This shrub grows in disturbed areas, rangeland, pastureland, drainage ditches and river bottoms, and along roadsides and fence lines.

Whole plant

Flower

Fruit

Stem

Leaf

Marshelder, narrowleaf sumpweed

Iva angustifolia DC.

Family: Asteraceae (Sunflower family)

Marshelder is a native, warm-season annual that is also commonly named narrowleaf sumpweed. A member of the Sunflower family, it is characterized by its leaflike bracts in the flowering stem. Marshelder is very drought tolerant.

This plant germinates in the early spring in February or March and is mostly vegetative, with long, narrow leaves. The flowers, which resemble those of the ragweed group, are inconspicuous. It flowers in late summer and fall.

Though seldom eaten, sumpweeds cause abortion in cattle. The pollen from this group is also a noted human allergen.

Distribution: Regions 1–7 and 10

Habitat: Marshelder occurs on seepy areas or those that may hold some water in the spring, especially along the edges of creeks and ponds.

Mohrs shin oak

Quercus mohriana Buckl. ex Rydb.
Family: Fagaceae (Beech family)

A shrubby oak in the Beech family, Mohrs shin oak usually grows as a short, thicket-forming shrub, seldom to tree form. The bark is grayish brown and is deeply furrowed on older stems.

The leaves are located alternately along the stems and are persistent (remaining attached beyond the usual time). Their oblong shape ends abruptly in a point. Although some leaves may have a few rounded lobes, most are not lobed. The acorns are borne annually on densely hairy peduncles (stalks that each bear one flower or fruit).

The acorns and young leaves of this shrub are toxic to grazing livestock, but the shrub provides good forage for browsing wildlife.

Distribution: Regions 7–10

Habitat: Mohrs shin oak is found on dry, well-drained limestone soils of west-central Texas.

Plantain

Plantago spp. L.
Family: Plantaginaceae (Plantain family)

Thirteen species of plantain are recognized in Texas. Most are native, cool-season annuals, but three species are perennials and two are introduced. The introduced species are common plantain and buckhorn, or English plantain. The most common plantains are redseed plantain, cedar or Heller plantain, and Hooker plantain.

Plantains are erect and stemless—all the leaves originate from a crown at the base of the plant. As winter annuals, the leaves lie flat on the ground in a rosette before spring growth.

The leaves of redseed plantain (*Plantago rhodosperma* Dcne.) vary from ⅓ inch to 2 inches wide. The leaves are pubescent, or lightly hairy, and the leaf margins can be toothed or narrow and elongated.

Cedar or Heller plantain (*Plantago hellerii* Small) is shorter than redseed but more villous, or very hairy, with narrow leaves. Hooker plantain is usually hairless or woolly. The leaves are narrow but can be of various lengths, depending on the growing conditions.

The flowering stems of plantains have spreading hairs, and the flowers have four parts. The fruit is a capsule with one to a few seeds.

Cool-season plantains can be of excellent forage value for livestock and wildlife in periods after drought.

Distribution: Regions 1–10

Habitat: Plantains are found on various disturbed soils of pastures, roadsides, parks, lawns and vacated areas.

Heller p. flower

Redseed p. whole plant

Redseed p. rosette

Heller p. whole plant

Post oak

Quercus stellata Wang.
Family: Fagaceae (Beech family)

A 30- to 60-foot-tall tree in the Beech family, post oak has a few large branches and a rounded crown. It is a perennial, warm-season native with reddish brown bark.

A post oak leaf is dark green, oblong, about 4 to 6 inches long and deeply five-lobed. The rounded middle lobes are located opposite each other, giving the leaf a cross-like appearance. The leaves are hairy underneath.

Post oak provides fair grazing for wildlife but can be poisonous to livestock.

Distribution: Regions 1–8

Habitat: The range of native post oaks extends from Central to East Texas.

Flower

Stem

Whole plant

Fruit

Leaf

Prairie gerardia, prairie agalinis

Agalinis heterophylla (Nutt.) Small ex Britt. & A. Br.
Family: Scrophulariaceae (Foxglove family)

Prairie gerardia is a native, warm-season annual in the Foxglove family. It is also commonly named prairie agalinis or grass killer, because it can shadow and suppress lower growing grass.

The plant grows to about 3 feet tall. A key characteristic of prairie gerardia is that its stems turn black when the plant dies in the summer or fall.

Depending on moisture, the purple tubular flowers appear in the summer, late summer and generally the fall.

The forage value of this plant is low for livestock and fair for wildlife.

Distribution: Regions 1–7

Habitat: Prairie gerardia is found mainly on prairies, plains, grasslands, open woodlands and fallow fields.

107

Whole plant

Flower

Fruit

Leaf

Pricklyash, tickle-tongue

Zanthoxylum hirsutum Buckl.
Family: Rutaceae (Citrus family)

Pricklyash is a shrub to small tree in the Citrus family. A native, cool-season perennial, it is also commonly named tickle-tongue or toothache tree. Pricklyash often occurs as a shrub in brushy areas and in sandy soils.

Pricklyash flowers in the spring and produces a small citrus fruit that has no pulp and one large, black seed.

Two other species of *Zanthoxylum* commonly occur in Texas. This species is common to Central Texas and the Hill Country. It is extremely common to find this plant growing along fence lines.

The forage value of pricklyash is poor for livestock and fair to poor for deer.

Distribution: Regions 1–8 and 10

Habitat: Pricklyash is found on sandy or gravelly soil of Central and West Texas.

Fruit

Seed

Whole plant

Stem

Leaf

Pricklypear

Opuntia spp. Mill.
Family: Cactaceae (Cactus family)

Pricklypears consist of jointed and flattened stems called pads or cladophylls. The plant's ability to store water in its flattened, fleshy stems enables it to withstand long dry or drought periods.

Although these highly modified stems are not leaves, they can conduct photosynthesis. True pricklypear leaves are tiny; they appear only briefly at each cluster of spines when new pads emerge.

It is often difficult to differentiate among species of pricklypear, mainly because of the way the plant reproduces. Each pad, if broken away, can form a new plant that is genetically identical to the original plant. Given enough time, an especially strong, adapted pricklypear can spread over a large area as its pads become scattered. Shredding and mowing pricklypear helps spread the plant and increase its density in a pasture.

Pricklypear fruits, or tunas, are eaten by people as well as cattle, sheep, goats, deer, javelina, turkeys and a wide assortment of other animals that process the pulp and pass the seeds through their digestive systems.

Distribution: Regions 1–10

Habitat: Pricklypear grows throughout Texas in a variety of habitats, from dry gravel or rocky, shallow soils to heavy, deep, clay soils. It is a common invader of grasslands and shrublands.

Population

Whole plant

Fruit

Flower

Rayless goldenrod, jimmyweed

Isocoma wrightii (Gray) Rydb.
Family: Asteraceae (Sunflower family)

Rayless goldenrod is a low-growing half-shrub in the Sunflower family. It has erect stems that arise from a woody crown and grow to a height of 2 to 4 feet.

The leaves are sticky, hairless, narrow and located alternately along the stems. The leaf margins may be even or slightly toothed. The stems bear flat-topped clusters of yellow flowers from June through October.

Rayless goldenrod is poisonous to livestock and provides poor forage for wildlife.

Distribution: Regions 8 and 10

Habitat: This plant is often found on dry rangelands, especially in river valleys, along drainage areas and irrigation canals, and on gypsiferous soil outcrops. It is a problem in the Pecos Valley drainage area in southeastern New Mexico and western Texas. It usually grows at elevations of 2,000 to 6,000 feet and is found from southern Colorado into Texas, Mexico, New Mexico and Arizona.

Flower

Whole plant

Stem

Leaf

Redberry juniper, Pinchot juniper

Juniperus pinchotii Sudw.
Family: Cupressaceae (Cypress family)

Redberry juniper is a perennial, cool- and warm-season native. This spreading, bushy tree in the Cypress family grows up to 10 feet tall and does not develop a central stem. Its bark is gray, thin and scalelike and peels off in narrow strips.

The leaves have resin-producing glands that produce white specks on the leaves. The leaves grow in dark green masses and are very slender, thin and sharp pointed. The fruit is red or reddish brown when mature.

This tree provides poor grazing for wildlife and livestock.

Distribution: Regions 5 and 7–10

Habitat: Redberry juniper is most common on dry rocky or gravelly soils, usually on open hills or flats. It is also found along creeks and drainage areas.

Population

Stem

Whole plant

Fruit

Leaf

Retama

Parkinsonia aculeata L.
Family: Fabaceae (Legume family)

Retama is a green-barked shrub or small tree in the Legume family. It grows 10 to 15 feet tall and has slender, spreading branches with feathery foliage. The trunks and branches are armed with needle-sharp spines that are turned up slightly.

Retama leaves are located alternately along the stems and are twice compound, with one or two branches and many leaflets per branch.

The yellow flowers have five petals, one of which has red spots. The fruit is a brown, many-seeded legume up to 4 inches long.

Retama leaves are sometimes browsed by white-tailed deer, and the fruit is eaten by deer and other mammals and birds.

Distribution: Regions 2–7 and 10

Habitat: Retama grows in moist, poorly drained areas.

Whole plant

Flower

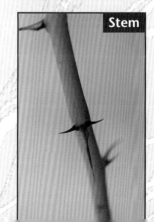
Stem

Leaf

Roosevelt willow

Baccharis neglecta Britt.
Family: Asteraceae (Sunflower family)

Roosevelt willow is a branching shrub in the Sunflower family. It grows 3 to 10 feet tall, its branches forming a near-rounded crown.

At the base of the plant, the leaves are broad and serrated on the outer margins; toward the flowering stem, the leaves narrow. The flowers appear from August to November in loose, diversely branching clusters. In the fall when they are mature, the flowers are feathery and showy. The fruits mature from November to December.

Roosevelt willow has low value for wildlife and livestock.

Distribution: Regions 2, 4, 6, 7 and 10

Habitat: This shrub is found in open woodlands and low prairies or along rivers and streams in the southern half and western parts of Texas.

Flower

Whole plant

Leaf

Flowering stem

Running live oak

Quercus virginiana Mill.
Family: Fagaceae (Beech family)

One of a group of live oak varieties in the Beech family, running live oak is characterized by short, scrubby growth and is a geographic variation. These trees often form large, dense thickets.

There are several subspecies of this tree, including var. *maritime*, which is commonly known as bay live oak and is common to sandy soils near the coast. Another subspecies, var. *fusiformis*, may occur as a shrubby variety in the Edwards Plateau region of Texas.

The bark may be dark brown to gray, depending on the subspecies. The leaves are simple and are located alternately along the stem. The leaves are evergreen and persistent (remaining attached to the stem through the winter).

The forage value of running live oak is fair for goats and wildlife.

Distribution: Regions 2, 3 and 5–7

Habitat: In Texas, this tree grows in sandy to shallow soils from the Rio Grande Plains through the Edwards Plateau.

Sacahuista

Nolina texana S. Wats.
Family: Liliaceae (Lily family)

Sacahuista is a perennial in the Lily family. It forms a large, distinctive clump of many fibrous, narrow leaves that can reach 5 feet long. The stems are woody and mostly buried.

When there is adequate rainfall, the plant gives rise in the spring to several stems bearing many clustered flowers. The flower stalks usually are not apparent until the plant is in full bloom.

The flowers and seeds of sacahuista are poisonous to livestock. The plant has poor forage value for most wildlife species.

Distribution: Regions 5–8 and 10

Habitat: Sacahuista is usually found on rocky range sites and mountain foothills from 3,000 to 7,000 feet in elevation. It is found in western Texas, Arizona, New Mexico and Mexico.

Flower

Whole plant

Fruit

Saltcedar, tamarisk

Tamarix spp. L.
Family: Tamaricaceae (Tamarisk family)

A native of Europe and Asia, saltcedar was introduced in the United States as an ornamental in the early 1800s. As its name implies, saltcedar can tolerate extreme salinity.

Most saltcedars are deciduous shrubs or small trees typically growing 10 to 30 feet tall and forming dense thickets. A few species are evergreen. The plant has slender branches and dense, gray-green foliage. The young twigs and stems have smooth, reddish brown bark.

The leaves are very small and scalelike, about $1/16$ inch long. They often have a crustlike scale from salt secretions. From March to September, the plant produces small white, pink or purple flowers in dense masses on its stem tips. It can produce up to 500,000 seeds per plant each year from April through October.

Saltcedar has fair forage value for goats and deer.

Distribution: Regions 1–3 and 5–10

Habitat: Saltcedar grows in moist soils near rivers, streams and lakes or in shallow groundwater tables.

Whole plant

Population

Flower, white

Flower, pink

Leaf

Seedling

Stem

Sand sagebrush

Artemisia filifolia Torr.
Family: Asteraceae (Sunflower family)

A small, aromatic shrub in the Sunflower family, sand sagebrush usually grows less than 3 feet tall. Its twigs are slender, dark gray to black and covered with fine, short hair.

The leaves are slender, long and light green. The flowers occur in dense, leafy clusters. The fruit is small, dry, hard, hairless and one-seeded, and it does not open at maturity.

Sand sagebrush can grow in densities high enough to retard grass growth.

The plant has low forage value for livestock and fair value for wildlife.

Distribution: Regions 7, 9 and 10

Habitat: Sand sagebrush is abundant in sandy soils to altitudes of 6,000 feet. It often grows in association with sand shinnery oak and small soapweed, as well as in pure stands.

127

Whole plant

Population

Leaf

Stem

Sand shinnery oak, Havard shin oak

Quercus havardii Rydb.
Family: Fagaceae (Beech family)

Sand shinnery oak is a low, shrubby tree of the Beech family. It rarely reaches more than 3 feet tall. Because of their aggressive underground rhizomes, these trees can form dense thickets over large areas and in deep sands.

The leaves are deciduous and located alternately along the stems. The leaves have a leathery, rough texture and pointed lobes.

Sand shinnery oaks produce typical oak-type acorn fruit. Produced annually in the spring, the acorns of this tree are rather large, ranging from ½ to 1 inch long and ½ to ⅝ inch wide.

The tree may hybridize with other shin oak and live oak species.

The young stems and acorns are poisonous to livestock and the plant provides fair forage for wildlife.

Distribution: Regions 7–10

Habitat: Sand shinnery oak grows in deep, sandy soils in the western part of Texas, including the lower Panhandle, Permian Basin and Trans-Pecos regions.

Population

Whole plant

Fruit

Leaf

Seepwillow

Baccharis salicifolia (R. & P.) Pers.
Family: Asteraceae (Sunflower family)

Seepwillow is a perennial, warm-season shrub of the Sunflower family. It grows along stream banks and can reach 3 to 12 feet tall. This shrub has wandlike stems that arise from a central base; the stems are flexible when young, becoming woody with maturity.

The leaves are willowlike, narrow at the base and tips, and broadly serrated along the margins. Its leaves are larger and broader than those of willow baccharis. Seepwillows bear small, fuzzy flowers at the ends of the stems from February through May.

The forage value is poor for livestock and fair for wildlife.

Distribution: Regions 2, 6, 7 and 10

Habitat: Seepwillows form thickets along streams and dry arroyos in West Texas.

Sericea lespedeza

Lespedeza cuneata (DuMont) G. Don
Family: Fabaceae (Legume family)

Sericea lespedeza is a shrubby, deciduous perennial of the Legume family. It grows 2 to 5 feet tall. A native of eastern Asia, it is widely cultivated and escaping throughout the southeastern United States.

The coarse stems of sericea lespedeza may grow singly or in clusters. New growth each year comes from buds on the stem bases about 1 to 3 inches below ground. The stems and branches are dense with wedge-shaped leaves, ¼ to 1 inch long and ¹⁄₁₆ to ¼ inch wide.

Each leaf is round at the top with a point at the tip. The lower leaf surface has silky hairs. The flowers are yellowish white with purple to pink markings; they appear from mid-July to early October.

The forage value is fair for livestock and good for wildlife. However, the plant can be invasive, eliminating or reducing native plants.

Distribution: Regions 1, 3 and 5

Habitat: Sericea lespedeza grows in ditches, along fence rows and on disturbed rangeland and pastures. It establishes best on disturbed areas with minimal vegetation.

Stem

Whole plant

Flower

Leaf

Silverleaf nightshade

Solanum elaeagnifolium Cav.
Family: Solanaceae (Nightshade family)

Silverleaf nightshade is an upright, usually prickly perennial in the Potato or Nightshade family. It normally grows 1 to 3 feet tall. This plant reproduces by seed and creeping root stalks.

Its characteristic silver color is imparted by the tiny, starlike, densely matted hairs covering the entire plant. The leaves have wavy margins and are lance shaped to narrowly oblong. The showy violet or bluish (sometimes white) flowers are followed by round, yellow fruits of up to ½ inch in diameter from May to October.

The plant has poor forage value for livestock and wildlife and can be poisonous to livestock.

Distribution: Regions 1–10

Habitat: Silverleaf nightshade is a serious weed of prairies, open woods and disturbed soils in southwestern United States and Mexico. It is occasionally found even farther north than Missouri.

Stem

Whole plant

Population

Leaf

Flower

Fruit

Skunkbush sumac, fragrant sumac

Rhus aromatica Ait.

Family: Anacardiaceae (Sumac family)

Skunkbush sumac is a perennial, warm-season native that has many branches. A low-growing shrub, it can reach 10 feet tall. Several subspecies of *R. aromatica* are recognized in Texas.

The scented leaves of this shrub are arranged in three leaflets. The flowers are yellow; the fruits are red.

Skunkbush sumac provides fair grazing for wildlife and livestock.

Distribution: Regions 1–5 and 7–10

Habitat: This shrub grows in a variety of habitats, ranging from woodlands, hills and sand dunes to rocky soils. It is generally confined to the eastern and west-central parts of Texas.

Stem

Whole plant

Leaf

Fruit

Smartweed

Polygonum pensylvanicum L.
Family: Polygonaceae (Knotweed family)

A member of the Knotweed family, smartweed is a native, warm-season annual. It is also commonly named Pennsylvania smartweed, pinkweed or pink smartweed.

The leaves are long and narrow, ending in a point. This plant bears white to pinkish flowers from March to November. It is commonly found growing around ponds and streams.

The forage value of smartweed is poor for livestock and fair for wildlife.

Distribution: Regions 2–9

Habitat: Smartweed grows in wet or disturbed soils.

Leaf

Whole plant

Flower

Stem

Spiny aster, wolfweed, Mexican devil-weed

Leucosyris spinosa (Benth.) Greene
Family: Asteraceae (Sunflower family)

Spiny aster is a native, warm-season perennial that can grow more than 3 feet tall. It is also called wolfweed, Mexican devil-weed or devilweed aster. It colonizes areas via underground rhizomes and can infest a large area.

The plant is characterized by green stems with stripes, occasional spines on the stems, and solitary sunflowers with white petals and small, yellow flowers in the center.

Small leaves are present for a brief period in the spring when the young stems are succulent. The leaves drop off after 2 to 3 weeks, and the stems begin conducting photosynthesis. The plant flowers in the summer and fall.

Because only some of the mature stems survive through the winter and continue growth and development the next season, mature stands of spiny aster often have an abundance of dead stems. A mature community may support more than 100 stems per square meter.

The forage value is poor for livestock and wildlife.

Distribution: Regions 1–10

Habitat: Spiny aster can grow along lowland streams and on roadsides, weedy slopes, stream banks, ditches, depressions, bottomlands, cultivated fields (chiefly cotton and alfalfa) and riverbanks. It invades sites with a high potential for forage production and upland sites where the soils have a high clay content and can hold much water.

Flower

Stem

Population

Whole plant

Leaf

Spiny hackberry, granjeno

Celtis pallida Torr.
Family: Ulmaceae (Elm family)

Spiny hackberry, or granjeno, is a native, cool-season perennial shrub reaching 4 to 15 feet tall. Growing along its stems are characteristic thorns that are sharp, heavy, unequal and paired. Its smooth, gray branches zigzag between the thorns.

The leaf blades are hairy and somewhat rough. Each leaf has three prominent veins and is toothed or sparsely toothed along the margins. The leaves are simple, arranged alternately along the stem, ½ to 2¼ inches long and ½ to 1 inch wide.

Spiny hackberry flowers are greenish white. The fruit is a yellow or orange—occasionally red—drupe, which is a fleshy fruit such as a peach or plum that usually contains a hard stone enclosing a seed.

This shrub provides excellent food and cover for wildlife. The fruit is eaten by a number of birds and mammals, and the leaves provide browse for white-tailed deer.

Distribution: Regions 2, 6 and 7

Habitat: Spiny hackberry is found on a variety of soils in brush thickets and other brushlands, mainly in South Texas and the Edwards Plateau.

Sulfaweed, broadleaf sumpweed

Iva annua L.

Family: Asteraceae (Sunflower family)

Sulfaweed is a native, warm-season annual in the Sunflower family. Also called broadleaf sumpweed, broadleaf marshelder, seacoast sumpweed and pelocote, the plant can grow from 2 to 6 feet tall. It has enough hair to be rough to the touch.

The leaves are situated in pairs across from each other on the stem. Each has three major veins visible on the lower or inner leaf.

Sulfaweed blooms in the summer and fall. A key characteristic is the presence of many small, leaflike bracts around every flower in the flowering stem.

The forage value of sulfaweed is fair for wildlife and, when the plant is young, fair for livestock.

Distribution: Regions 1–9

Habitat: This plant grows in moist soils in disturbed areas, roadsides and coastal bermudagrass pasture.

Sweetgum

Liquidambar styraciflua L.
Family: Hamamelidaceae (Witch-hazel family)

Sweetgum is a large, native tree in the Witch-hazel family. A cool-season perennial, it can grow to 150 feet tall. Its bark is brown to gray and very rough and may have deep furrows and rounded ridges.

The leaves are simple, deciduous and arranged alternately on the stem. Each leaf is attached to the twig by a petiole (leaf stem) and has three to seven (usually five) lobes that taper to a point. The leaves are glossy above, but underneath they have fine, short hairs along the veins.

The greenish flowers have no showy petals; they occur on 2- to 3-inch-long flowering stems. Each flower grows on its own stalk along the stem. The fruit is an aggregate of many two-celled, globe-shaped ovaries. The fruit are green before maturing and then turn brown.

Sweetgum provides food for about 25 species of birds that feed on the fruit. It offers little value as a grazing plant.

Distribution: Regions 1–3

Habitat: Sweetgum grows in low bottomlands and in moist to dry uplands.

Stem

Leaf

Whole plant

Fruit

Tarbush

Flourensia cernua DC.
Family: Asteraceae (Sunflower family)

Tarbush is a strongly aromatic, perennial shrub of the Sunflower family. Its stems are brittle and have a characteristic black color.

The plant's height may vary from 1 foot tall on dry sites to more than 6 feet tall in deep, overflow areas. It is most common on deep soils. The roots are shallow to take advantage of the quick thunderstorms common to the desert region where it grows.

The leaves of tarbush are smooth along the edges and oval or oblong. They are located alternately along the stems. The flowers are solitary in the leaf axils (the angle between the upper leaf and the leaf stem), forming a leafy flowering stem in the fall.

The forage value of tarbush is poor for livestock and wildlife.

Distribution: Regions 7, 8 and 10

Habitat: Tarbush is common on dry plains, hills and mesas from counties just east of the Pecos River in Texas, west to Arizona and south to Mexico.

149

Population

Whole plant

Leaf

Stem

Flower

Tasajillo, turkey pear

Opuntia leptocaulis DC.
Family: Cactaceae (Cactus family)

Tasajillo has cylindric, jointed stems in which the individual segments are ½ inch or less in length. A member of the Cactus family, the plant has a general brushy appearance and usually grows to less than 5 feet tall.

Tasajillo leaves are tiny; they appear only briefly at each cluster of spines when new segments emerge. The plant produces small, inconspicuous, greenish flowers. The fruit is red, fleshy, globe shaped and less than an inch long. The spines are white, slender and from ¾ inch to 2 inches long.

Tasajillo often reaches the greatest densities along fence lines or under trees, where the seeds are transported by birds.

This plant spreads by seeds, which are relished by wildlife.

Distribution: Regions 2 and 4–10

Habitat: Tasajillo is most abundant on sandy loam and clay loam soils and is often found in association with honey mesquite.

Fruit

Whole plant

Leaf

Stem

Texas bullnettle, mala mujer

Cnidoscolus texanus (Muell. Arg.) Small
Family: Euphorbiaceae (Spurge family)

Texas bullnettle is a spiny, deep-rooted, herbaceous perennial in the Spurge family. It usually grows 1 to 3 feet tall. The plant may have one to several spiny stems from a single root system. If a stem is broken, a milky white sap appears.

The leaves of bullnettle are dark green and located alternately along the stems. Each leaf is divided into five leaflets that are crinkled in appearance and covered with stiff, stinging hairs.

The flowers, which appear from April through September, consist of five to seven white, showy, petal-like sepals that are united below. The seeds are compartmentalized in a three- to four-celled fruit with a tough outer shell.

The forage value of bullnettle is poor for livestock and wildlife.

Distribution: Regions 1–8 and 10

Habitat: Bullnettle is a very common plant found in all soil types across Texas. It is most common on sandy soils and disturbed areas and thrives in hot weather. It is a very aggressive competitor in improved pastures.

Whole plant

Stem

Flower

Fruit

Leaf

Texas persimmon

Diospyros texana Scheele
Family: Ebenaceae (Persimmon family)

Texas persimmon, also called Mexican or black persimmon, is a shrub or small tree that normally grows to less than 15 feet tall. However, some specimens along the upper Texas coast may reach 50 feet tall. Its compact wood is almost black, and the bark is gray, slick and thin.

Texas persimmon has oval leaves that are rounded at the tips. The leaves have small, fine hairs underneath. The fruit contains three to eight seeds and can measure up to 1 inch in diameter. It is green at first, turning black when ripe.

The forage value of Texas persimmon is fair for goats and wildlife.

Distribution: Regions 2–10

Habitat: Texas persimmon is found primarily in the western two-thirds of the state in rocky, open woodlands, arroyos, and on open slopes. In some pastures in Central Texas, it may be one of the predominant invading woody species.

Flower

Whole plant

Fruit

Leaf

Thistles

Cirsium spp. Mill.
Family: Asteraceae (Sunflower family)

Ten species of thistles are found in Texas. One of the more common species, Texas thistle (*C. texanum*), is a prickly, tap-rooted annual that reaches 3 to 5 feet tall.

Each plant begins growth as a rosette, or a circular cluster of leaves, during the winter and sends up a flowering stalk in the spring. The leaves are extremely spiny and located alternately along the stems. In some species, the leaf bottom may be woolly with small hairs.

Showy purple flowers occur in late spring and summer and mature into a white plume. The seeds scatter profusely in the wind.

Although the forage value of thistles is poor for livestock and wildlife, seed-eating birds feed on the seeds of many thistles.

Distribution: Regions 1–10

Habitat: Thistles are common in dry or moist soils throughout Texas. They thrive in disturbed or overgrazed areas, in abandoned fields and along roadsides.

Whole plant

Leaf

Flower

Threadleaf groundsel

Senecio douglasii DC.
Family: Asteraceae (Sunflower family)

Threadleaf groundsel is a many-stemmed evergreen composite in the Sunflower family. The stems are herbaceous, although somewhat woody at the base, and may have variable hairiness.

The stems and leaves are gray-green. The leaves are long, threadlike and divided into three to seven segments. They may be hairy or nearly smooth. Showy yellow flowers emerge from March through November.

Threadleaf groundsel is poisonous to livestock and offers poor to fair forage value for wildlife.

Distribution: Regions 6–10

Habitat: Threadleaf groundsel is a common range plant in Colorado and Utah and south to Texas and Mexico. It is common in the grassland areas of western Texas. Disturbance and overgrazing can cause it to increase in abundance.

Stem

Whole plant

Flower

Leaf

Twinleaf senna, twoleaf senna

Senna roemariana (Scheele) Irwin & Barneby
Family: Fabaceae (Legume family)

Twinleaf senna is an erect, perennial herb in the Legume family. The plant is green and covered with short, soft hairs, which give it a grayish appearance. Twinleaf senna can have a few or many stems arising from a thickened root.

The leaves are arranged spirally as pairs of leaflets (hence the name twinleaf). The pairs of leaflets fold together during dry periods to conserve moisture.

Deep yellow flowers emerge from April to August; each has five petals that are about twice as long as the sepals, which are the usually green flower parts that surround and protect the bud. The flower stamens, or pollen-bearing organs, are straw-colored to light brown. A bean-type fruit develops after flowering.

This plant is poisonous to goats.

Distribution: Regions 4–10

Habitat: Twinleaf senna is common in pastures and open woods on limestone soils in Central and West Texas and westward to New Mexico.

Flower

Whole plant

Fruit

Leaf

Twisted acacia, huisachillo

Acacia schaffneri (S. Wats) Herm.
Family: Fabaceae (Legume family)

Twisted acacia is a spiny, spreading, multi-stemmed shrub of the Legume family. It can reach 4 to 12 feet tall. Its stems have many spines that are paired, pin-like and pale or blackish.

The leaves are twice compound, and the flowers are round and yellowish to orange.

Twisted acacia is sometimes confused with huisache but can be distinguished from huisache by its round growth habit, longer and narrower legumes, and the petiolar (leaf stem) gland located between the lowermost pair of leaf branches.

Several species of wildlife make use of twisted acacia. It is browsed by white-tailed deer, and the fruit is eaten by javelina, feral hogs and some birds. It is also used for loafing, nesting and protective cover by birds and small mammals.

Distribution: Regions 2, 6 and 7

Habitat: Twisted acacia grows in various soils in mixed-brush stands and root-plowed areas.

Whole plant

Fruit

Flower

Leaf

Stem

Upright prairie coneflower

Ratibida columnifera (Nutt.) Woot. & Stndl.
Family: Asteraceae (Sunflower family)

Upright prairie coneflower is a native, warm-season perennial in the Sunflower family. Also called Mexican hat, it is a common weed and wildflower of roadsides, parks, vacated lands and managed pastures.

This hairy-stemmed plant reproduces from seed or short underground stems. The stems grow from 12 to 40 inches tall and branch near the top. The leaves are strongly lobed into distinct segments that are long, narrow and pointed.

The flowers are borne at the end of slender stems. Each has yellow to brownish petals and a dark brown center that can reach 1 inch long.

Upright prairie coneflower provides good grazing for wildlife and poor grazing for livestock.

Distribution: Regions 1–10

Habitat: This plant commonly grows along roadsides and in parks, vacated lands and managed pastures.

Flower variation

Flower

Whole plant

Leaf

Western bitterweed

Hymenoxys odorata DC.
Family: Asteraceae (Sunflower family)

Western bitterweed is an erect, annual, composite plant in the Sunflower family. It reaches 3 inches to 2 feet tall. The stems are purplish near the base. This plant has a bitter taste and a distinct odor.

The leaves usually are woolly underneath and are located alternately along the stems.

Bright yellow flowers bloom from April through June and occasionally in the fall. With adequate rainfall, it is common to see pastures turn yellow with this plant in the spring.

Western bitterweed is highly toxic to sheep, especially during drought.

Distribution: Regions 3–10

Habitat: Bitterweed is common in arid areas of the southern Great Plains from southwestern Kansas and central Texas to southern California and into Mexico. It is most common where soil disturbance or overgrazing has occurred. The populations can vary considerably from year to year.

Flower

Whole plant

Leaf

Population

Western honey mesquite

Prosopis glandulosa Torr. var. *torreyana* (Benson) M.C. Johnst.
Family: Fabaceae (Legume family)

This variety of common honey mesquite is found in the Trans-Pecos region of Texas. A member of the Legume family, western honey mesquite generally reaches 2 to 4 feet tall and is multi-stemmed. The two varieties of *Prosopis glandulosa* hybridize readily, and the plants may contain characteristics of each.

The leaves are composed of five to 20 pairs of leaflets that are much smaller than those of common honey mesquite. They are opposite, or situated in pairs across from each other along the stem. The plant also has opposite thorns at each node.

Its flowers can appear throughout the summer months, depending on rainfall. The long, narrow bean pods develop and fall to the ground during late summer to fall.

The forage value of western honey mesquite is poor for livestock and wildlife except that the beans are consumed by both.

Distribution: Regions 7 and 10

Habitat: Western honey mesquite grows in dry native ranges mainly west of the Pecos River in Texas.

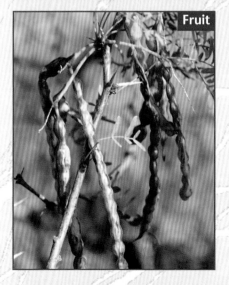

Western horse nettle, treadsalve

Solanum dimidiatum Raf.
Family: Solanaceae (Nightshade family)

Western horse nettle is a weedy perennial in the Nightshade family. It is also called treadsalve. Although most specimens of this species are less than 2 feet tall, some grow to 3 feet.

The leaves, petioles (leaf stems) and branched stems of western horse nettle carry sharp spines. Western horse nettle leaves are oval, have five to seven lobes and reach up to 6 inches long.

The flowers are bluish purple to violet (rarely white) and grow in clusters at the ends of the stems. They give rise to spherical fruits that are ¾ to 1½ inches in diameter and yellow at maturity.

The fruit of the plant is poisonous to livestock, and the forage value is poor for grazing wildlife.

Distribution: Regions 2–8

Habitat: These plants are widely distributed on loamy prairies of Texas except in the far western, Panhandle and extreme eastern parts of the state.

Whole plant

Flower

Leaf

Fruit

Western ragweed

Ambrosia cumanensis Kunth in H.B.K.
Family: Asteraceae (Sunflower family)

Western ragweed is a native, warm-season perennial in the Sunflower family. A common Texas plant, it is also called perennial ragweed.

Western ragweed is rhizomatous and can reproduce through vegetative tillers that come from the parent plant. The plant grows from long rootstock, with stout, many-branched stems that can reach heights of 12 to 72 inches.

The leaves are thick, gray-green and hairy or bristly. They are 2 to 5 inches long and have deep serrations along the margins. The serrations are sometimes pointed, sometimes rounded. The plant flowers from late summer through fall.

The seeds of western ragweed are borne along a central stem that grows 2 to 6 inches long at the top of the plant. The seed clusters are chaffy (like grain husks), becoming pointed and bristly with maturity.

This plant provides good grazing for wildlife and poor grazing for livestock.

Distribution: Regions 1–10

Habitat: Western ragweed grows mainly on disturbed sites, dry rangelands and hillsides. It is a common rangeland weed across Texas.

Whitebrush, beebush

Aloysia gratissima (Gill. & Hook.) Troncoso
Family: Verbenaceae (Vervain family)

Whitebrush is an aromatic shrub of the Vervain family. It reaches 3 to 10 feet tall. Also called beebush, this shrub may grow as a single plant or in large, dense thickets.

The leaves are narrow, small, pointed, pale beneath and ¼ to 1 inch long. On the flowering branches, the leaves are smaller and smooth-edged; those on other branches are larger and toothed. The tiny flowers vary from white to blue.

Whitebrush is poisonous to horses, mules and burros but provides fair browse for wildlife.

Distribution: Regions 1–7 and 10

Habitat: Whitebrush is frequent to abundant in Central, West and South Texas. It grows northwest into New Mexico and south into Mexico. In far West Texas, this plant is usually restricted to draws receiving extra runoff moisture and having deep soils.

Whole plant

Stem

Flower

Leaf

White-thorn acacia, mescat acacia

Acacia constricta Gray
Family: Fabaceae (Legume family)

A member of the Legume family, white-thorn acacia has typical legume-type leaves made up of three to nine pairs of leaflets. The leaflets may be located alternately along the stem or situated in pairs across from each other on the stem.

The stems are generally dark brown or black with characteristic paired, straight, white spines up to 1½ inches long.

Its showy yellow flowers are spherical and very aromatic. The flowering date varies considerably with moisture conditions. The fruit of white-thorn acacia is a long, slender, linear legume or bean pod, usually maturing from July to September.

This plant has no forage value for livestock or wildlife.

Distribution: Regions 6, 7 and 10

Habitat: White-thorn acacia grows in dry sandy, caliche or rocky soils in the western and mountain regions of Texas.

Flower

Fruit

Whole plant

Stem

Leaf

Wild carrot

Daucus pusillus Michx.
Family: Apiaceae (Parsley family)

Wild carrot is an introduced, cool-season annual that is also commonly named southwestern carrot or rattlesnakeweed. It grows as an erect, single-stemmed plant reaching 1 to 3 feet tall. When crushed, the taproot has the characteristic odor of a carrot.

The leaves and stems are covered with stiff hairs, making the plant rough to the touch. The leaves are divided pinnately, having leaflets arranged on each side of the stalk, and can be from 1 to 7 inches long.

The flowering stem is a flat-topped cluster of white flowers in which each flower stalk arises from about the same point. When the fruit ripen, they may cling to passing animals or the clothing of people for seed dispersal.

Wild carrot has little to no value for grazing livestock or wildlife.

Distribution: Regions 1–10

Habitat: This plant grows along roadsides and in fields, over-grazed pastures and disturbed areas throughout the state.

Willow

Salix spp. L.
Family: Salicaceae (Willow family)

More than 80 species and varieties of *Salix* grow in Texas. Willows are deciduous trees or shrubs that form large, dense root mats on the soil surface or in shallow water and slow-moving streams.

The leaves of most species are long and narrow, with finely toothed edges. The undersides of the leaves are a lighter color than the upper. The flower stalks carry many tiny flowers.

Willows have small seeds, each with long, silky hairs attached to one end like a parachute, which helps them spread. The seeds usually live only a few days to a few weeks.

The forage value of willows is generally poor for wildlife and livestock.

Distribution: Regions 1–10

Habitat: Willows are found on dry areas in their eastern range and on moist soils along streams in their western range.

Willow baccharis

Baccharis salicina T. & G.
Family: Asteraceae (Sunflower family)

Willow baccharis is a smooth shrub in the Sunflower family. A perennial, warm-season native, it has many branches and grows to 3 or more feet tall.

The leaves are long, narrow, dotted with resin and located alternately along the stems. They may be entire or sparingly indented on the margin. The leaves taper at the base and are fairly pointed at the tip.

Willow baccharis blooms in late summer and fall. The flowers occur in clusters on short flower stalks and form a single series of dull, white bristles at maturity.

This shrub provides little to no value to grazing livestock and wildlife.

Distribution: Regions 5, 8 and 10

Habitat: Willow baccharis is found mainly in moist soils in open areas and often in disturbed areas or along roadsides.

183

Leaf

Flower

Stem

Whole plant

Winged elm

Ulmus alata Michx.
Family: Ulmaceae (Elm family)

Winged elm is a native, cool-season perennial tree that can reach 60 feet tall. This deciduous plant is most easily recognized by the corky wings on its branches and stems. The twigs and bark are reddish brown to gray brown.

The leaves of winged elm are simple and arranged alternately on the stems. They grow from ⅓ inch to 3 inches long. The leaf margin is doubly serrated.

The flowers of winged elm occur from February to April before the leaves appear. The fruit, which is winged for easy dispersal by wind, ripens in March and April.

The leaves are a medium-preferred browse of grazing animals, and the fruit is eaten by seed-eating birds.

Distribution: Regions 1–5 and 7

Habitat: Winged elm grows on a variety of soils along streams and in woodlands and thickets. This species also occurs as scattered trees in native and introduced pastures.

Winged stem

Fruit

Whole plant

Mature stem

Leaf

Woolly locoweed

Astragalus mollissimus Torr.
Family: Fabaceae (Legume family)

Woolly locoweed is a stout, many-branched perennial of the Legume family. Its thick, woody root gives rise to stems lying close to the ground.

The leaves of woolly locoweed have 19 to 29 oval to oblong leaflets that are covered with fine, soft, short hairs. The flowers are purple, lavender or yellow, emerging from April through June. They mature into thick, inflated, moon-shaped seedpods.

This plant is poisonous to all species of livestock and wildlife.

Distribution: Regions 7–10

Habitat: Woolly locoweed grows from southwestern South Dakota to Texas and New Mexico. It is common in upland, mesa and mountain areas of the Trans-Pecos and Panhandle regions of Texas.

Yankeeweed, rosinweed

Eupatorium compositifolium Walt.
Family: Asteraceae (Sunflower family)

Yankeeweed is an upright, mainly single-stemmed perennial in the Sunflower family. A warm-season native, it is produced from strong underground rootstock and has stems ranging from 24 to 48 inches tall.

The lower leaves are lobed, compound and located alternately on the stems; the upper leaves are less compound and often entire. Its flowers are white and produced as a long head.

This plant provides poor grazing for wildlife and livestock.

Distribution: Regions 1–3 and 6

Habitat: Yankeeweed grows in disturbed areas, abandoned fields and roadsides. It prefers sandy soils.

Leaf

Whole plant

Flower

Yaupon

Ilex vomitoria Soland. in Ait.
Family: Aquifolaceae (Holly family)

Yaupon is a low-growing, evergreen shrub in the Holly family. It will form a trunk when trimmed but usually grows as a woody shrub. A native, cool- and warm-season perennial, this plant has branches that can reach 20 feet tall.

The leaves are oval, 2 to 4 inches long and entire with toothed margins. The flowers are white, and the fruits at maturity are bright red.

Yaupon provides good grazing for wildlife, fair grazing for livestock.

Distribution: Regions 1–4, 6 and 7

Habitat: This shrub grows in moist, acidic soils in the eastern part of the state.

Yucca

Yucca spp. L.
Family: Liliaceae (Lily family)

Texas has more than 30 species of yucca, many cultivated as ornamentals. They are members of the Lily family. Yuccas are usually large plants that have either a distinct, woody trunk above ground or thick, branching, mostly underground main stems.

The leaves are numerous from the base, commonly long, narrow and spine-tipped. They can be thin and flaccid (limp) or thick and rigid.

The usually large and numerous flowers emerge along a tall flowering stem. They are white to cream colored or greenish. The fruits can be dry or fleshy, with many seeds in each cell. The fruit are flattened and usually blackish.

The forage value of yucca is poor for livestock; some species offer fair forage value for wildlife.

Distribution: Regions 1–10

Habitat: This plant is most common in arid regions.

Flower

Whole plant, fruit

Stem

Whole plant, flower

Range Plants by Region

Region 1: annual broomweed, annual croton: oneseed, annual croton: woolly, bitter sneezeweed, blackberry, blackjack oak, bois d'arc, buffalo-bur, camphorweed, Carolina horse nettle, Chinese tallow tree, cocklebur, common persimmon, common sunflower, curlycup gumweed, eastern baccharis, eastern red cedar, elm, flameleaf sumac, greenbriar, hackberry, Hercules club, honey locust, honey mesquite, horehound, Macartney rose, marshelder, plantain, post oak, prairie gerardia, pricklyash, pricklypear, saltcedar, sericea lespedeza, silverleaf nightshade, skunkbush sumac, spiny aster, sulfaweed, sweetgum, Texas bullnettle, thistles, upright prairie coneflower, western ragweed, whitebrush, wild carrot, willow, winged elm, yankeeweed, yaupon, yucca

Region 2: annual broomweed, annual croton: oneseed, annual croton: Texas, annual croton: woolly, Berlandier lobelia, bitter sneezeweed, blackberry, blackbrush, blackjack oak, bois d'arc, broom snakeweed, buffalo-bur, camphorweed, Carolina horse nettle, catclaw acacia, ceniza, Chinese tallow tree, cocklebur, Drummonds goldenweed, common persimmon, common sunflower, eastern baccharis, eastern red cedar, elm, flameleaf sumac, gray goldaster, greenbriar, guajillo, hackberry, Hercules club, honey locust, honey mesquite, horehound, huisache, lime pricklyash, lotebush, Macartney rose, marshelder, plantain, post oak, prairie gerardia, pricklyash, pricklypear, retama, Roosevelt willow, running live oak, saltcedar, seepwillow, silverleaf nightshade, skunkbush sumac, smartweed, spiny aster, spiny hackberry, sweetgum, sulfaweed, tasajillo, Texas bullnettle, Texas persimmon, thistles, twisted acacia, upright prairie coneflower, western horse nettle, western ragweed, whitebrush, wild carrot, willow, winged elm, yankeeweed, yaupon, yucca

Region 3: annual broomweed, annual croton: oneseed, annual croton: Texas, annual croton: woolly, bitter sneezeweed, blackberry, blackjack oak, bois d'arc, broom snakeweed, buffalo-bur, camphorweed, Carolina horse nettle, cocklebur, common persimmon, common sunflower, curlycup gumweed, eastern baccharis, eastern red cedar, elm, flameleaf sumac, greenbriar, hackberry, Hercules club, honey locust, honey mesquite, horehound, huisache, lotebush, Macartney rose, marshelder, plantain, post oak, prairie gerardia, pricklyash, pricklypear, retama, running live oak, saltcedar, sericea lespedeza, silverleaf nightshade, skunkbush sumac, smartweed, spiny aster, sweetgum, sulfaweed, Texas bullnettle, Texas persimmon, thistles, upright prairie coneflower, western bitterweed, western horse nettle, western ragweed, whitebrush, wild carrot, willow, winged elm, yankeeweed, yaupon, yucca

Region 4: annual broomweed, annual croton: oneseed, annual croton: Texas, annual croton: woolly, bitter sneezeweed, blackberry, blackjack oak, bois d'arc, broom snakeweed, buffalo-bur, camphorweed, Carolina horse nettle, cocklebur, common persimmon, common sunflower, curlycup gumweed, eastern red cedar, elm, flameleaf sumac, gray goldaster, greenbriar, hackberry, Hercules club, honey locust, honey mesquite, horehound, huisache, Macartney rose, marshelder, lotebush, plantain, post oak, pricklyash, prairie gerardia, pricklypear, retama, Roosevelt willow, silverleaf nightshade, skunkbush sumac, smartweed, spiny aster, sulfaweed, tasajillo, Texas bullnettle, Texas persimmon, thistles, twinleaf senna, upright prairie coneflower, western bitterweed, western horse nettle, western ragweed, whitebrush, wild carrot, willow, winged elm, yaupon, yucca

Region 5: annual broomweed, annual croton: oneseed, annual croton: Texas, annual croton: woolly, ashe juniper, bitter sneezeweed, blackberry, bois d'arc, broom snakeweed, buffalo-bur, camphorweed, catclaw acacia, catclaw mimosa, cocklebur, common persimmon, common sunflower, curlycup gumweed, dog cactus, elm, flameleaf sumac, gray goldaster, greenbriar, hackberry, honey locust, honey mesquite, horehound, lotebush, Macartney rose, marshelder, plantain, post oak, prairie gerardia, pricklyash, pricklypear, redberry juniper, retama, running live oak, sacahuista, saltcedar, sericea lespedeza, silverleaf nightshade, skunkbush sumac, smartweed, spiny aster, sulfaweed, tasajillo, Texas bullnettle, Texas persimmon, thistles, twinleaf senna, upright prairie coneflower, western bitterweed, western horse nettle, western ragweed, whitebrush, wild carrot, willow, willow baccharis, winged elm, yucca

Region 6: annual broomweed, annual croton: oneseed, annual croton: Texas, annual croton: woolly, ashe juniper, Berlandier lobelia, bitter sneezeweed, blackbrush, broom snakeweed, buffalo-bur, camphorweed, catclaw acacia, ceniza, cocklebur, common goldenweed, Drummond's goldenweed, common sunflower, creosotebush, dog cactus, elm, gray goldaster, greenbriar, guajillo, hackberry, honey locust, honey mesquite, horehound, huisache, lime pricklyash, lotebush, marshelder, plantain, post oak, prairie gerardia, pricklyash, pricklypear, retama, Roosevelt willow, running live oak, sacahuista, saltcedar, seepwillow, silverleaf nightshade, smartweed, spiny aster, spiny hackberry, sulfaweed, tasajillo, Texas bullnettle, Texas persimmon, thistles, threadleaf groundsel, twinleaf senna, twisted acacia, upright prairie coneflower, western bitterweed, western horse nettle, western ragweed, whitebrush, white-thorn acacia, wild carrot, willow, yankeeweed, yaupon, yucca

Region 7: African rue, annual broomweed, annual croton: one-seed, annual croton: Texas, annual croton: woolly, ashe juniper, Bigelow shinoak, bittersneezeweed, blackberry, blackbrush, blackjack oak, bois d'arc, broom snakeweed, buffalo-bur, burrobrush, camphorweed, Carolina horse nettle, catclaw acacia, catclaw mimosa, ceniza, cholla, Christ thorn, cocklebur, common persimmon, common sunflower, creosotebush, curlycup gumweed, dog cactus, eastern red cedar, elm, flameleaf sumac, gray goldaster, greenbriar, guajillo, hackberry, Hercules club, honey locust, honey mesquite, horehound, huisache, lotebush, Macartney rose, marshelder, Mohrs shin oak, plantain, post oak, prairie gerardia, pricklyash, pricklypear, redberry juniper, retama, Roosevelt willow, running live oak, sacahuista, saltcedar, sand sagebrush, sand shinnery oak, seepwillow, silverleaf nightshade, skunkbush sumac, smartweed, spiny aster, spiny hackberry, sulfaweed, tarbush, tasajillo, Texas bullnettle, Texas persimmon, thistles, threadleaf groundsel, twinleaf senna, twisted acacia, upright prairie coneflower, western bitterweed, western honey mesquite, western horse nettle, western ragweed, whitebrush, white-thorn acacia, wild carrot, willow, winged elm, woolly locoweed, yaupon, yucca

Region 8: African rue, annual broomweed, annual croton: one-seed, annual croton: Texas, annual croton: woolly, ashe juniper, bitter sneezeweed, blackberry, blackjack oak, bois d'arc, broom snakeweed, buffalo-bur, camphorweed, catclaw acacia, catclaw mimosa, cholla, cocklebur, common persimmon, common sunflower, curlycup gumweed, elm, gray goldaster, greenbriar, hackberry, honey mesquite, horehound, lotebush, Mohrs shin oak, plantain, post oak, pricklyash, pricklypear, rayless goldenrod, redberry juniper, sacahuista, saltcedar, sand shinnery oak, silverleaf nightshade, skunkbush sumac, smartweed, spiny aster, sulfaweed, tarbush, tasajillo, Texas bullnettle, Texas persimmon, thistles, threadleaf groundsel, twinleaf senna, upright prairie coneflower, western bitterweed, western horse nettle, western ragweed, wild carrot, willow, willow baccharis, woolly locoweed, yucca

Region 9: annual broomweed, annual croton: Texas, annual croton: woolly, ashe juniper, bitter sneezeweed, broom snakeweed, buffalo-bur, camphorweed, catclaw mimosa, cholla, cocklebur, common sunflower, curlycup gumweed, garboncillo, gray goldaster, hackberry, honey mesquite, horehound, lotebush, Mohrs shin oak, plantain, pricklypear, redberry juniper, saltcedar, sand sagebrush, sand shinnery oak, silverleaf nightshade, skunkbush sumac, smartweed, spiny aster, sulfaweed, tasajillo, Texas persimmon, thistles, threadleaf groundsel, twinleaf senna, upright prairie coneflower, western bitterweed, western ragweed, wild carrot, willow, woolly locoweed, yucca

Region 10: African rue, annual broomweed, annual croton: one-seed, annual croton: Texas, annual croton: woolly, ashe juniper, bitter sneezeweed, blackbrush, bois d'arc, broom snakeweed, buffalo-bur, burrobrush, camphorweed, catclaw acacia, catclaw mimosa, ceniza, cholla, cocklebur, common sunflower, creosotebush, curlycup gumweed, dog cactus, elm, garboncillo, gray goldaster, guajillo, hackberry, honey mesquite, horehound, lotebush, marshelder, Mohrs shin oak, plantain, pricklyash, pricklypear, rayless goldenrod, redberry juniper, retama, Roosevelt willow, sacahuista, saltcedar, sand sagebrush, sand shinnery oak, seepwillow, silverleaf nightshade, skunkbush sumac, spiny aster, tarbush, tasajillo, Texas bullnettle, Texas persimmon, thistles, threadleaf groundsel, twinleaf senna, upright prairie coneflower, western bitterweed, western honey mesquite, western ragweed, whitebrush, whitethorn acacia, wild carrot, willow, willow baccharis, woolly locoweed, yucca

Index

A

Anacardiaceae, 70, 136
Acacia berlandieri Benth., 78
Acacia constricta Gray, 176
Acacia greggii Gray, 38
Acacia: mescat, 176; sweet, 90; twisted, 162; whitethorn, 176
Acacia rigidula Benth., 22
Acacia schaffneri (S. Wats) Herm., 162
Acacia smallii Isely, 90
African rue, 2
Agalinis heterophylla (Nutt.) Small ex Britt. & A. Br., 106
agalinis, prairie, 106
Aloysia gratissima (Gill. & Hook.) Troncoso, 174
Ambrosia cumanensis Kunth in H.B.K., 172
Amphiachyris dracunculoides (DC.) Nutt., 4
annual broomweed, 4
annual croton: one-seed, 6; Texas, 8; woolly, 10
annual sunflower, 56
Apiaceae, 178
Aquifolaceae, 190
Artemisia filifolia Torr., 126
Ashe juniper, 12
Asteraceae, 4, 18, 28, 32, 34, 50, 52, 56, 60, 64, 74, 98, 112, 118, 126, 130, 140, 144, 148, 156, 158, 164, 166, 172, 182, 188
aster: devilweed, 140; spiny, 140
Astragalus mollissimus Torr., 180
Astragalus wootonii Sheldon, 72

B

baccharis: eastern, 64; willow, 182
Baccharis halimifolia L., 64
Baccharis neglecta Britt., 118
Baccharis salicifolia (R. & P.) Pers., 130
Baccharis salicina T. & G., 182
basin sneezeweed, 18
bay live oak, 120
beebush, 174
Beech family, 16, 24, 100, 104, 120, 128
Berlandier lobelia, 14
Bigelow shinoak, 16
bitter sneezeweed, 18
bitterweed, western, 166
blackberry, 20
blackbrush, 22
blackjack oak, 24
black persimmon, 154
Bluebell family, 14
blue brush, 94
blueberry juniper, 12
bois d'arc, 26
broadleaf marshelder, 144
broadleaf sumpweed, 144
broom snakeweed, 28
broomweed: annual, 4; common, 4; perennial, 28
Buckthorn family, 48, 94
buffalo-bur, 30
bullnettle, Texas, 152
burrobrush, 32

C

Cactaceae, 46, 62, 110, 150
cactus, dog, 62
Cactus family, 46, 62, 110, 150
Caltrop family, 2, 58
camphorweed, 34
Campanulaceae, 14
Carolina horse nettle, 36
carrot, southwestern, 178
catclaw acacia, 38
catclaw mimosa, 40
cedar, 12; eastern red, 66
cedar elm, 68
cedar plantain, 102
Celtis pallida Torr., 142
Celtis spp. L., 80
ceniza, 42
cheese brush, 32
cheesewood, 32
Chinese tallow tree, 44
cholla, 46; dog, 62; walking stick, 46
Christ thorn, 48
Cirsium spp. Mill., 156
Citrus family, 82, 92, 108
clavellina, 62
clepe, 94
Cnidoscolus texanus (Muell. Arg.) Small, 152
cocklebur, 50
colima, 92
common broomweed, 4
common goldenweed, 52
common persimmon, 54
common sunflower, 56
coneflower, upright prairie, 164
consumption-weed, 64
creosotebush, 58
Croton capitatus Michx., 10
Croton monanthogynus Michx., 6
Croton texensis (Klotzch) Muell. Arg., 8
Croton: one-seed, 6; Texas, 8; woolly, 10
Cupressaceae, 12, 66, 114
curlycup gumweed, 60
Cypress family, 12, 66, 114

D

Daucus pusillus Michx., 178
devil-weed, Mexican, 140
devilweed, aster, 140
dewberry, 20
Diospyros texana Scheele, 154
Diospyros virginiana L., 54
dog cactus, 62
dog cholla, 62
dog pear, 62
Drummonds goldenweed, 52

E

eastern baccharis, 64
eastern persimmon, 54
eastern red cedar, 66
Ebenaceae, 54, 154
elm, 68; American, 68; cedar, 68; slippery, 68; winged, 184

Elm family, 68, 80, 142, 184
Eupatorium compositifolium Walt., 188
Euphorbiaceae, 6, 8, 10, 44, 152

F

Fabaceae, 22, 38, 40, 72, 78, 84, 86, 90, 116, 132, 160, 162, 168, 176, 186
Fagaceae, 16, 24, 100, 104, 120, 128
flameleaf sumac, 70
Flourensia cernua DC., 148
Foxglove family, 42, 106
fragrant mimosa, 40
fragrant sumac, 136

G

garboncillo, 72
gerardia, prairie, 106
Gleditsia triacanthos L., 84
goldaster, gray, 74
goldenweed: common, 52; Drummonds, 52
goldenrod, rayless, 112
granjeno, 142
gray goldaster, 74
greasewood, 58
greenbriar, 76; saw, 76
Grindelia squarrosa (Pursh) Dun., 60
guajillo, 78
gumdrop tree, 94
gumweed, curlycup, 60
Gutierrezia sarothrae, (Pursh.) Britt. & Rusby, G. microcephala (DC.) Gray, 28

H

hackberry, 80; netleaf, 80; spiny, 142; sugar, 80; western, 80
Hamamelidaceae, 146
hat, Mexican, 164
Havard shin oak, 128
Helenium amarum (Raf.) H. Rock var. *amarum*, 18
Helenium amarum (Raf.) H. Rock var. *badium* (Gray ex S. Wats.) Waterfall, 18
Helianthus annuus L., 56
Heller plantain, 102
Hercules club, 82
Heterotheca canescens (DC.) Shinners, 74
Heterotheca subaxillaris (Lam.) Britt. & Rusby var. *latifolia* (Buckl.) Gandhi & Thomas, 34
Holly family, 190
honey locust, 84
honey mesquite, 86; western, 168
Hooker plantain, 102
horehound, 88
horse nettle: Carolina, 36; western, 170
huisache, 90
huisachillo, 162
Hymenoclea monogyra T. & G. ex Gray, 32
Hymenoxys odorata DC., 166

I

Ilex vomitoria Soland. in Ait., 190
Iscoma coronopifolia (Gray) Greene, 52
Iscoma drummondii (T. & G.) Greene, 52
Isocoma wrightii (Gray) Rydb., 112
Iva angustifolia DC., 98
Iva annua L., 144

J

Jerusalem thorn, 48
jimmyweed, 112
juniper: Ashe, 12; blueberry, 12; Pinchot, 114; redberry, 114
Juniperus ashei Buchholz, 12
Juniperus pinchotii Sudw., 114
Juniperus virginiana L., 66

K

Knotweed family, 138

L

Lamiaceae, 88
Larrea tridentata (DC.) Cov., 58
Legume family, 22, 38, 40, 72, 78, 84, 86, 90, 116, 132, 160, 162, 168, 176, 186
Lespedeza cuneata (Du-Mont) G. Don, 132
lespedeza, sericea, 132
Leucophyllum frutescens (Berl.) I.M. Johnst., 42
Leucosyris spinosa (Benth.) Greene, 140
Liliaceae, 76, 122, 192
Lily family, 76, 122, 192
lime pricklyash, 92
Liquidambar styraciflua L., 146
live oak: bay, 120; running, 120
lobelia, Berlandier, 14
Lobelia berlandieri A. DC., 14
locoweed, woolly, 186
locust, honey, 84
lotebush, 94

M

Macartney rose, 96
Maclura pomifera (Raf.) Schn., 26
mala mujer, 152
marshelder, 98; broadleaf, 144
Marubium vulgare L., 88
mescat acacia, 176
mesquite: honey, 86; western honey, 168
Mexican devil-weed, 140
Mexican hat, 164
Mexican persimmon, 154
mimosa: catclaw, 40; fragrant, 40
Mimosa biuncifera Benth., 40
Mimosa borealis Gray, 40
Mint family, 88
Mohrs shin oak, 100
Moraceae, 26
Mulberry family, 26

N

narrowleaf sumpweed, 98
netleaf hackberry, 80
Nightshade family, 30, 36, 134, 170
nightshade, silverleaf, 134
Nolina texana S. Wats., 122

O

oak: bay live, 120; blackjack 24; Havard shin, 128; Mohrs shin, 100; post, 104; running live, 120; sand shinnery, 128; scalybark, 16; scrub, 16; shin, 16
Opuntia imbricata (Haw.) DC., 46
Opuntia leptocaulis DC., 150
Opuntia schottii Engelm., 62
Opuntia spp. Mill., 110
osage orange, 26

P

Paliurus spina-christi Mill., 48
Parkinsonia aculeata L., 116
Parsley family, 178
pear: dog, 62 turkey, 150
Peganum harmala L., 2
pelocate, 144
Pennsylvania smartweed, 138
perennial broomweed, 28
perennial ragweed, 172
persimmon: black, 154; common, 54; eastern, 54; Mexican, 154; Texas, 154
Persimmon family, 54, 154
pink smartweed, 138
pinkweed, 138
Pinchot juniper, 114
Plantaginaceae, 102
Plantago spp. L., 102
plantain, 102: cedar, 102; Heller, 102; Hooker, 102; redseed, 102
Plantain family, 102
Polygonaceae, 138
Polygonum pensylvanicum L., 138
post oak, 104
prairie agalinis, 106
prairie coneflower, upright, 164

prairie gerardia, 106
pricklyash, 108; lime, 92
pricklypear, 110
Prosopis glandulosa Torr. var. *glandulosa*, 86
Prosopis glandulosa Torr. var. *torreyana* (Benson) M.C. Johnst., 168
purplesage, 42

Q

Quercus durandii Buckl. var. *breviloba* (Torr.) Palmer, 16
Quercus havardii Rydb., 128
Quercus marilandica Muenchh., 24
Quercus mohriana Buckl. ex Rydb., 100
Quercus stellata Wang., 104
Quercus virginiana Mill., 120

R

ragweed: perennial, 172; western, 172
Ratibida columnifera (Nutt.) Woot. & Stndl., 164
rattlesnake weed, 72
rattleweed, 72
rayless goldenrod, 112
redberry juniper, 114
red cedar, eastern, 66
redseed plantain, 102
retama, 116
Rhamnaceae, 48, 94
Rhus aromatica Ait., 136
Rhus copallina L., 70
Roosevelt willow, 118
Rosa bracteata Wendl., 96
Rosaceae, 20, 96
Rose family, 20, 96
rose, Macartney, 96
rosinweed, 188
Rubus spp., 20
rue, African, 2
running live oak, 120
Rutaceae, 82, 92, 108

S

sacahuista, 122
sagebrush, sand, 126
Salicaceae, 180
Salix spp. L., 180
saltcedar, 124
sand sagebrush, 126

sand shinnery oak, 128
Sapium sebiferum (L.) Roxb., 44
saw greenbriar, 76
scalybark oak, 16
Scrophulariaceae, 42, 106
scrub oak, 16
seacoast sumpweed, 144
seepwillow, 130
Senecio douglasii DC., 158
Senna roemariana (Scheele) Irwin & Barneby, 160
senna: twinleaf, 160; two-leaf, 160
sericea lespedeza, 132
shinnery oak, sand, 128
shinoak: Bigelow, 16; white, 16
shin oak, 16: Havard, 128; Mohrs, 100
silverleaf nightshade, 134
silverleaf, Texas, 42
skunkbush sumac, 136
smartweed, 138; Pennsylvania, 138; pink, 138
Smilax bona-nox L., 76
snakeweed, broom, 28
sneezeweed: basin, 18; bitter, 18
Solanaceae, 30, 36, 134, 170
Solanum carolinense L., 36
Solanum dimidiatum Raf., 170
Solanum elaeagnifolium Cav., 134
Solanum rostratum Dun., 30
southwestern carrot, 178
spiny aster, 140
spiny hackberry, 142
Spurge family, 6, 8, 10, 44, 152
sugarberry, 80
sugar hackberry, 80
sulfaweed, 144
Sumac family, 70, 136
sumac: flameleaf, 70; fragrant, 136; skunkbush, 136
sumpweed: broadleaf, 144; narrowleaf, 98; seacoast, 144
sunflower: annual, 56; common, 56
Sunflower family, 4, 18, 28, 32, 34, 50, 52, 56, 60,

64, 74, 98, 112, 118, 126, 130, 140, 144, 148, 156, 158, 164, 166, 172, 182, 188
sweet acacia, 90
sweetgum, 146

T
Tamaricaceae, 124
tamarisk, 124
Tamarisk family, 124
Tamarix spp. L., 124
tarbush, 148
tasajillo, 150
Texas bullnettle, 152
Texas persimmon, 154
Texas silverleaf, 42
thistles, 156
threadleaf groundsel, 158
tickle-tongue, 108
toothache tree, 82
treadsalve, 170
turkey pear, 150
twinleaf senna, 160
twisted acacia, 162
twoleaf senna, 160

U
Ulmaceae, 68, 80, 142, 184
Ulmus alata Michx., 184
Ulmus spp. L., 68
upright prairie coneflower, 164

V
Verbenaceae, 174
Vervain family, 174

W
walking stick cholla, 46
western bitterweed, 166
western hackberry, 80
western honey mesquite, 168
western horse nettle, 170
western ragweed, 172
whitebrush, 174
white shinoak, 16
white-thorn acacia, 176
wild carrot, 178
willow, 180; Roosevelt, 118
willow baccharis, 182
Willow family, 180
winged elm, 184
Witch-hazel family, 146
wolfweed, 140
woolly croton, 10
woolly locoweed, 186

X
Xanthium strumarium L., 50

Y
yankeeweed, 188
yaupon, 190
yucca, 192
Yucca spp. L., 192

Z
Zanthoxylum clava-herculis L., 82
Zanthoxylum fagara (L.) Sarg., 92
Zanthoxylum hirsutum Buckl., 108
Zizyphus obtusifolia (T. & G.) Gray var. obtusifolia, 94
Zygophyllaceae, 2, 58

Produced by AgriLife Communications, the Texas A&M System
Extension publications can be found on the Web at
http://agrilifebookstore.org

Visit theTexas AgriLife Extension Service at
http://AgriLifeExtension.tamu.edu

Educational programs of the Texas AgriLife Extension Service are open to all people without regard to race, color, sex, disability, religion, age or national origin.

Issued in furtherance of Cooperative Extension Work in Agriculture and Home Economics, Acts of Congress of May 8, 1914, as amended, and June 30, 1914, in cooperation with the United States Department of Agriculture. Edward G. Smith, Director, the Texas AgriLife Extension Service, the Texas A&M System.

ISBN 0-9721049-4-1